ALSO BY EMILY ANTHES

Instant Egghead Guide: The Mind

FRANKENSTEIN'S CAT

FRANKENSTEIN'S CAT

Cuddling Up to Biotech's Brave New Beasts

EMILY ANTHES

SCIENTIFIC AMERICAN / FARRAR, STRAUS AND GIROUX
NEW YORK

Scientific American / Farrar, Straus and Giroux
18 West 18th Street, New York 10011

Printed in the United States of America
First edition, 2013

An excerpt from *Frankenstein's Cat* originally appeared, in slightly different form, in *Scientific American*.

Library of Congress Cataloging-in-Publication Data
Anthes, Emily.
Frankenstein's cat : cuddling up to biotech's brave new beasts / Emily Anthes. — First edition.
pages cm
Includes bibliographical references and index.
ISBN 978-0-374-15859-0 (hardcover : alk. paper)
1. Transgenic animals. I. Title.

QH442.6 .A58 2013
616.02'73—dc23

2012029045

Designed by Abby Kagan
Illustrations by Diego Patino

www.fsgbooks.com • books.scientificamerican.com
www.twitter.com/fsgbooks • www.facebook.com/fsgbooks

10 9 8 7 6 5 4 3 2 1

For my family—humans and canines alike

Contents

FRANKENSTEIN'S CAT

Introduction

In China, the world's manufacturing powerhouse, a new industry is taking shape: the mass production of mutant mice. Peek into the 45,000 mouse cages at Shanghai's Fudan University and you'll see a growing collection of misfits. By randomly disabling the rodents' genes, the scientists here are churning out hundreds of odd animals, assembly-line style. They have created mice studded with skin tumors and mice that grow tusks. There's a mouse with male-pattern baldness, hair everywhere save for a lonely bare spot on its head. Some of the mice have strange behavioral quirks—they endlessly bury marbles, for instance, or make only left turns. One strain ages at warp speed. Another can't feel pain.

While some of the rodents have obvious abnormalities, others reveal their secrets over time. One variety appears normal on the outside, with thick white fur and healthy pink ears and noses. But the animals are klutzes. They are clumsy and spectacularly uncoordinated. They fail miserably when researchers put them through their paces at a special rodent boot camp. In one test, the mice are tasked with standing on top of a rotating rod for as long as they can manage, the rodent equivalent of a log-rolling challenge. It's not an easy undertaking, but normal mice eventually find their footing. The mutant mice never do. They also have trouble balancing on a narrow wooden beam and keeping their grip when suspended, upside down, from a wire screen. And they have strange gaits—taking abnormally wide steps and holding their tails at odd angles, curved up toward the ceiling, instead of letting them simply drag along the floor behind them, as mice usually do.

Even stranger, perhaps, are the Lonely Hearts Club mice. The males of this strain look like regular rodents, but the females consistently refuse to mate with them. The poor guys, lacking some certain je ne sais quoi, simply have no sex appeal, and they are rejected time and time again.

These mice are just a small sample of the more than 500 different kinds of mutants the Fudan team has created. Ultimately, the researchers hope to create *100,000* strains of modified mice, each eccentric in its own way. It would be enough to fill a carnival sideshow thousands of times over.

As long as we're dreaming up animal sideshows, we needn't stop with peculiar mice. Science has given us a whole new toolbox for tinkering with life, and we have the power to modify animals in profound new ways. We are editing their genetic codes, rebuilding their broken bodies, and supplementing their natural senses. Headlines frequently herald the birth of strange new creatures: *Bionic beetles! Glowing cats! Spider goats! Roborats!* The breakthroughs are simultaneously astounding and puzzling. What *are* these creatures exactly? What do they look like? Who's creating them, and why? And are these animals really so novel?

Indeed, we have a long history of refashioning animal bodies. Take the varied members of the species *Canis lupus familiaris*—the modern dog—which are products of millennia of life with humans and bear little resemblance to their ancestors, gray wolves. Exactly how this dog domestication began is a subject of intense debate. Some scientists suggest that we deliberately set out to acquire canine companions, adopting wild wolf pups. Others hypothesize that hungry wolves, attracted to the bones, trash, and scraps produced by early humans, approached our camps on their own terms, and that our tolerance of the least threatening interlopers gave rise to future generations of human-friendly canines. Either way, as wolves became part of human society, moving from cold ground to warm hearth, they lost many of the traits they needed to survive in the wild. Their bodies and heads shrank, their faces and jaws grew more compact, and their teeth decreased in size.

As our relationship with canines developed, we began to breed them more carefully, molding dogs that excelled at specific tasks. We created the bulky, barrel-chested mastiff to guard our homes, and the dachshund, a wiggly salami of a dog, to shimmy into badger burrows. The diversity among modern dogs is so astounding that the thirty thousand dogs that strut their canine stuff at Crufts, the largest dog show in the world, don't even look like members of the same species. One year, the "Best in Show" contenders included King, a hound with a deer's build, all legs and lean muscle, and Ricky, a tiny black-and-white fluff ball who could stand easily underneath King's smooth brown belly. They shared the ring with Donny—a standard poodle whose shaved gray haunches were set off by a thick white mane—and Cruella, an Old English sheepdog whose long, shaggy hair obscured all but the black dot that presumably served as her nose. Today, thanks to us, dogs are the most physically diverse species on Earth.

We've reshaped other species, too, turning scrawny chickens into plump broiler birds and bristly-haired wild sheep into producers of soft wool. The list goes on and on. We learned to breed animals that suited our every need, creating hunters, herders, guardians, food sources, and companions. Over the course of generations, the members of many species diverged from their wild ancestors and took their place in a human world.

But selective breeding was a blunt instrument, one that required us to transform animals using educated guesswork, breeding desirable hounds together, over and over again, until a puppy we liked squirmed into the world. It took thousands of years to turn wolves into dogs. Now we can create novel organisms in years, months, even days.

Today, the tools of molecular biology allow us to target one specific gene, to instantly turn it on or off, to silence or amplify its effects. For instance, the researchers at Fudan University are creating their stunning array of strange mice simply by knocking out a single gene at a time. To do so, they're relying on a special genetic tool called a transposon or a "jumping gene," a segment of DNA capable of hopping around the genome. When the scientists inject a transposon into a mouse embryo, this foreign piece of DNA inserts itself into a random place in the rodent's genome, disabling whatever gene it finds there. But the real beauty of the system is that when this mouse grows up and mates, the transposon jumps to a different location in the genome of its pups, sabotaging a new gene. With each mating, researchers have no idea where the transposon will end up, what gene it will disrupt, or what the ultimate effects will be. It's like throwing darts at a genetic dartboard. Blindfolded. Only when the pups are born, and start exhibiting various abnormalities, do the scientists learn what part of the genome has gone haywire. The approach is allowing the researchers to create cages upon cages of novel mutants, simply by playing matchmaker between their amorous rodents. In some cases, the scientists are making furry freaks faster than they can figure out what's wrong with them.

We can also recombine genes in ways that nature never would—just consider a very curious cat skulking about New Orleans. With downy orange fur and a soft pink nose, the feline looks like your average tabby. But flick on a black light, and the cat becomes Mr. Green Genes, his nose turning from soft pink to electric lime, due to a bit of jellyfish DNA tucked into each of his cells. The insides of his ears and the whites of his eyes glow brightly, his face emerging from the dark like a modern-day Cheshire cat. (His son, Kermit, also glows green.)

Meanwhile, nearly two thousand miles away, a barn in Logan, Utah, is home to a strange herd of goats. Thanks to a pair of genes borrowed from a spider, each female goat produces milk that's chock-full of silk proteins. When the milk is processed in the lab, scientists can extract the spider proteins and spin them into silk.

Genetics isn't the only field providing us with the power to reengineer other species. Advances in electronics and computing make it possible to merge animal bodies with machines, to use tiny electrodes to hijack a rat's brain and guide the rodent, like a remote-controlled toy, through a complicated obstacle course. Breakthroughs in materials science and veterinary surgery are helping us build bionic limbs for injured animals, and we can train monkeys to control robotic arms with their thoughts. Today, our grandest science fiction fantasies are becoming reality.

Some of us may find our growing control over living, breathing beings to be unsettling. After all, biotechnology is the stuff of dystopian nightmares, and many an apocalyptic scenario has been constructed around crazy chimeras or world-conquering cyborgs. Ethicists and activists worry about whether we should be altering other species when we can't possibly get their consent. Some say that manipulating the planet's wild things—whether we're inserting genes or electrodes—is profoundly unnatural, causes animal suffering, and turns other life-forms into commodities. Critics worry that our effort to remake the world's fauna is the worst example of human hubris, the expression of an arrogant desire to play God.

It's true that remaking other species according to our own wants and needs doesn't necessarily put animal welfare first. Selective breeding hasn't always turned out well for animals—we've saddled dog breeds with all sorts of hereditary diseases and created turkeys with such gigantic breasts that they can barely walk. And of course, biotechnology gives us new ways to do damage. The Fudan University scientists have created mouse embryos with defects so severe that they die in the womb. Some of their mutant mice are prone to tumors, or kidney disease, or neurological problems. One strain, unable to absorb nutrients from food, essentially starves to death.

In fact, a whole industry has sprung up to sell diseased lab animals to scientists, with numerous biotech companies hawking their unique creations. In October 2011, many of these companies converged on St. Pete Beach, Florida, for an international meeting of scientists who work with genetically modified organisms. Representatives from various biotech firms held court from booths ringing a hotel ballroom, advertising animals that had been engineered to suffer from all sorts of medical afflictions. One company was selling pigs with cystic fibrosis and cancer; a brochure from another outlined eleven available strains of rodents, from the NSE-p25 mouse, designed to display Alzheimer's-like symptoms, to the 11BHSD2 mouse, which has a tendency to drop dead of heart failure. (And just in case nothing there caught your fancy, one company's poster promised, "You design the experiment, we'll design the mice.") These companies aren't making sickly animals purely to be cruel, of course; studying these creatures yields valuable insight into human disease. That's good news for us, but little consolation for a tumor-riddled rodent.

If there is peril here, there is also great promise. Biotechnology could do more for animals than it's given credit for. Sure, we can make animals sick, but we can also choose to deploy our species-shaping powers to help other species survive and thrive, to create healthier, happier, fitter critters, and some scientists are doing just that. With the sophisticated techniques at our fingertips, we may even be able to undo some of the damage we've done to other species, alleviating genetic disorders in dogs, for instance, or bringing wild animal populations back from the brink of extinction. Some forward-thinking philosophers are dreaming of more extreme interventions, such as boosting the brainpower of apes, and using genetic modification and electronic enhancement to help animals transcend the limits of their own bodies.

Right now all the options are open. Though biotechnology's strange new creatures are being created in the world's labs, they don't tend to stay there very long, and there are already cutting-edge animals living in fields, homes, and nature preserves across America. Before long, we may all be able to shop for animals the same way that scientists in Florida shopped for carefully engineered mice. Imagine a future in which we can each pick out the perfect animal from a catalogue of endless options. We could create something for everyone. Avid nighttime reader? How about your own Mr. Green Genes so you can stay up late, reading by the light of the cat? For the twelve-year-old who has everything, skip the toy cars and planes at Christmas and wrap up a remote-controlled rodent. Equestrians could order up a foal with the same genes as the winner of last year's Kentucky Derby, while sprinters could get themselves a golden retriever whose artificial carbon-fiber legs would allow it to run as fast as a greyhound. The tools of biotechnology are becoming increasingly accessible to the public; future generations of animal lovers may be able to design their own creatures without fancy lab equipment or advanced scientific training.

In the pages that follow, we'll go on a journey from petri dish to pet store, seeking out the revolutionary breeds of beasts that are taking their places in the world. We'll venture from the rocky shores of California to the dusty fields of Texas, from the canine clones that live in Korean labs to the pets that sleep in our homes. We'll delve into genes and brains, into work that seems frivolous and projects that are anything but. We'll meet an engineer who is turning beetles into stunt planes and a biologist who believes cloning just might save endangered species. And, of course, we'll come to know the animals themselves—from Jonathan, a sad sack of a seal with hundreds of online friends, to Artemis, a potentially life-saving goat whose descendants could one day take over Brazil.

Along the way, we'll puzzle through some larger questions. We'll probe how our contemporary scientific techniques are different from what's come before and whether they represent a fundamental change in our relationship with other species. We'll consider the relationship we have with animals and the one we'd like to have.

Most of us care deeply about some form of animal life, whether it's the cat or dog curled up on the couch—60 percent of Americans share their homes with pets of one species or another—the chickens laying our eggs, or some exotic predator fighting to survive as its habitat disappears. Now that we can sculpt life into an endless parade of forms, what we choose to create reveals what it is we want from other species—and what we want *for* them. But even if you feel no special affection for the creatures with whom we share this planet, our reinvention of animals matters for us, too. It provides a peek into our own future, at the ways we may start to enhance and alter ourselves. Most of all, our grand experiments reveal how entangled the lives of human and nonhuman animals have become, how intertwined our fates are. Enterprising scientists, entrepreneurs, and philosophers are dreaming up all sorts of projects that could alter the course of our collective future.

So what does biotechnology really mean for the world's wild things? And what do our brave new beasts say about us? Our search for answers begins with a tank of glowing fish.

1. Go Fish

To an aspiring animal owner, Petco presents an embarrassment of riches. Here, in the basement of a New York City store—where the air carries the sharp tang of hay and the dull musk of rodent dander—is a squeaking, squealing, almost endless menagerie of potential pets. There are the spindly-legged lizards scuttling across their sand-filled tanks; the preening cockatiels, a spray of golden feathers atop their heads; and, of course, the cages of pink-nosed white mice training for a wheel-running marathon. There are chinchillas and canaries, dwarf hamsters, tree frogs, bearded dragons, red-footed tortoises, red-bellied parrots, and African fat-tailed geckoes.

But one of these animals is not like the others. The discerning pet owner in search of something new and different merely has to head to the aquatic display and keep walking past the speckled koi and fantail bettas, the crowds of goldfish and minnows. And there they are, cruising around a small tank hidden beneath the stairs: inch-long candy-colored fish in shades of cherry, lime, and tangerine. Technically, they are zebrafish (*Danio rerio*), which are native to South Asian lakes and rivers and usually covered with black and white stripes. But these swimmers are adulterated with a smidgen of something extra. The Starfire Red fish contain a dash of DNA from the sea anemone; the Electric Green, Sunburst Orange, Cosmic Blue, and Galactic Purple strains all have a nip of sea coral. These borrowed genes turn the zebrafish fluorescent, so under black or blue lights they glow. These are GloFish, America's first genetically engineered pets.

Though we've meddled with many species through selective breeding, these fish mark the beginning of a new era, one in which we have the power to directly manipulate the biological codes of our animal friends. Our new molecular techniques change the game. They allow us to modify species quickly, rather than over the course of generations; doctor a single gene instead of worrying about the whole animal; and create beings that would never exist in nature, mixing and matching DNA from multiple species into one great living mash-up. We have long desired creature companions tailored to our *exact* specifications. Science is finally making that precision possible.

Though our ancestors knew enough about heredity to breed better working animals, our ability to tinker with genes directly is relatively new. After all, it wasn't until 1944 that scientists identified DNA as the molecule of biological inheritance, and 1953 that Watson and Crick deduced DNA's double helical structure. Further experiments through the '50s and '60s revealed how genes work inside a cell. For all its seeming mystery, DNA has a straightforward job: It tells the body to make proteins. A strand of DNA is composed of individual units called nucleotides, strung together like pearls on a necklace. There are four distinct types of nucleotides, each containing a different chemical base. Technically, the bases are called adenine, thymine, guanine, and cytosine, but they usually go by their initials: A, T, C, and G. What we call a "gene" is merely a long sequence of these As, Ts, Cs, and Gs. The order in which these letters appear tells the body which proteins to make—and where and when to make them. Change some of the letters and you can alter protein manufacturing and the ultimate characteristics of an organism.

Once we cracked the genetic code, it wasn't long before we figured out how to manipulate it. In the 1970s, scientists set out to determine whether it was possible to transfer genes from one species into another. They isolated small stretches of DNA from *Staphylococcus*—the bacteria that cause staph infections—and the African clawed frog. Then they inserted these bits of biological code into *E. coli*. The staph and frog genes were fully functional in their new cellular homes, making *E. coli* the world's first genetically engineered organism. Mice were up next, and in the early 1980s, two labs reported that they'd created rodents carrying genes from viruses and rabbits. Animals such as these mice, which contain a foreign piece of DNA in their genomes, are known as transgenic, and the added genetic sequence is called a transgene.

Encouraged and inspired by these successes, scientists started moving DNA all around the animal kingdom, swapping genes among all sorts of swimming, slithering, and scurrying creatures. Researchers embarking on these experiments had multiple goals in mind. For starters, they simply wanted to see what was possible. How far could they push these genetic exchanges? What could they *do* with these bits and pieces of DNA?

There was also immense potential for basic research; taking a gene from one animal and putting it into another could help researchers learn more about how it worked and the role it played in development or disease. Finally, there were promising commercial applications, an opportunity to engineer animals whose bodies produced highly desired proteins or creatures with economically valuable traits. (In one early project, for instance, researchers set out to make a leaner, faster-growing pig.)

Along the way, geneticists developed some neat tricks, including figuring out how to engineer animals that glowed. They knew that some species, such as the crystal jellyfish, had evolved this talent on their own. One moment, the jellyfish is an unremarkable transparent blob; the next it's a neon-green orb floating in a dark sea. The secret to this light show is a compound called green fluorescent protein (GFP), naturally produced by the jellyfish, which takes in blue light and reemits it in a kiwi-colored hue. Hit the jelly with a beam of blue light, and a ring of green dots will suddenly appear around its bell-shaped body, not unlike a string of Christmas lights wrapped around a tree.

When scientists discovered GFP, they began to wonder what would happen if they took this jellyfish gene and popped it into another animal. Researchers isolated and copied the jellyfish's GFP gene in the lab in the 1990s, and then the real fun began. When they transferred the gene into roundworms, rats, and rabbits, these animals also started producing the protein, and if you blasted them with blue light, they also gave off a green glow. For that reason alone, GFP became a valuable tool for geneticists. Researchers testing a new method of genetic modification can practice with GFP, splicing the gene into an organism's genome. If the animal lights up, it's obvious that the procedure worked. GFP can also be coupled with another gene, allowing scientists to determine whether the gene in question is active. (A green glow means the paired gene is on.)

Scientists discovered other potential uses, too. Zhiyuan Gong, a biologist at the National University of Singapore, wanted to use GFP to turn fish into living pollution detectors, swimming canaries in underwater coal mines. He hoped to create transgenic fish that would blink on and off in the presence of toxins, turning bright green when they were swimming in contaminated water. The first step was simply to make fish that glowed. His team accomplished that feat in 1999 with the help of a common genetic procedure called microinjection. Using a tiny needle, he squirted the GFP gene directly into some zebrafish embryos. In some of the embryos, this foreign bit of biological code managed to sneak into the genome, and the fish gave off that telltale green light. In subsequent research, the biologists also made strains in red—thanks to a fluorescent protein from a relative of the sea anemone—and yellow, and experimented with adding these proteins in combination. One of their published papers showcases a neon rainbow of fish that would do Crayola proud.*

To Richard Crockett, the co-founder of the company that sells GloFish, such creatures have more than mere scientific value—they have an obvious aesthetic beauty. Crockett vividly remembers learning about GFP in a biology class. He was captivated by an image of brain cells glowing green and red, thanks to the addition of the genes for GFP and a red fluorescent protein. Crockett was a premed student, but he was also an entrepreneur. In 1998, at the age of twenty-one, he and a childhood friend, Alan Blake, launched an online education company. By 2000, the company had become a casualty of the dot-com crash. As the two young men cast about for new business ideas, Crockett thought back to the luminescent brain cells and put a proposal to Blake: What if they brought the beauty of fluorescence genes to the public by selling glowing, genetically modified fish?

* In 2005, Gong's team announced that they had successfully used GFP to create medaka—another species of small fish native to Asia—that did indeed turn green when they were exposed to environmental estrogens, synthetic chemicals that can disrupt the hormones of humans and other animals. In 2010, scientists at China's Fudan University achieved a similar breakthrough with zebrafish. Despite these advances, South Korea, host of the 2010 G20 Summit, took a far cruder approach when it employed a school of security fish to protect the world's leaders from contaminated water: If the goldfish swimming around in tanks of the water died, well, that might indicate a problem.

At first, Blake, who had no background in science, thought his friend was joking. But when he discovered that Gong and other scientists were already fiddling with fish, he realized that the idea wasn't far-fetched at all. Blake and Crockett wouldn't even need to invent a new organism—they'd just need to take the shimmering schools of transgenic fish out of the lab and into our home tanks.

The pair founded Yorktown Technologies to do just that, and Blake took the lead during the firm's early years, setting up shop in Austin, Texas. He licensed the rights to produce the fish from Gong's lab and hired two commercial fish farms to breed the pets. (Since the animals pass their fluorescence genes on to their offspring, all Blake needed to create an entire line of neon pets was a few starter adults.) He and his partner dubbed them GloFish, though the animals aren't technically glow-in-the-dark—at least, not the same way that a set of solar system stickers in a child's bedroom might be. Those stickers, and most other glow-in-the-dark toys, work through a scientific property known as phosphorescence. They absorb and store light, reemitting it gradually over time, as a soft glow that's visible when you turn out all the lights. GloFish, on the other hand, are fluorescent, which means that they absorb light from the environment and beam it back out into the world immediately. The fish appear to glow in a dark room if they're under a blue or black light, but they can't store light for later—turn the artificial light off, and the fish stop shining.

Blake was optimistic about their prospects. As he explains, "The ornamental fish industry is about new and different and exciting varieties of fish." And if new, different, and exciting is what you're after, what more could you ask for than an animal engineered to glow electric red, orange, green, blue, or purple thanks to a dab of foreign DNA? Pets are products, after all, subject to the same marketplace forces as toys or clothes. Whether it's a puppy or a pair of heels, we're constantly searching for the next big thing. Consider the recent enthusiasm for "teacup pigs"—tiny swine cute enough to make you swear off pork chops forever.

Harold Herzog, a psychologist at Western Carolina University who specializes in human-animal interactions, has studied the way our taste in animals changes over time. When Herzog consulted the registry of the American Kennel Club, he found that dog breed choices fade in and out of fashion the same way that baby names do. One minute, everyone is buying Irish setters, naming their daughters Heather, and listening to "Bennie and the Jets"—welcome to 1974!—and then it's on to the next great trend. Herzog discovered that between 1946 and 2003, eight breeds—Afghan hounds, chow chows, Dalmatians, Dobermans, Great Danes, Old English sheepdogs, rottweilers, and Irish setters—went through particularly pronounced boom and bust cycles. Registrations for these canines would skyrocket, and then, as soon as they reached a certain threshold of popularity, people would begin searching for the next fur-covered fad.

Herzog identified a modern manifestation of our long-standing interest in new and unusual animals. In antiquity, explorers hunted for far-flung exotic species, which royal households often imported and displayed. Even the humble goldfish began as a luxury for the privileged classes. Native to Central and East Asia, the wild fish are usually covered in silvery gray scales. But ancient Chinese mariners had noticed the occasional yellow or orange variant wriggling in the water. Rich and powerful Chinese families collected these mutants in private ponds, and by the thirteenth century, fish keepers were breeding these dazzlers together. Goldfish domestication was born, and the once-peculiar golden fish gradually spread to the homes of less-fortunate Chinese families—and households elsewhere in Asia, Europe, and beyond.

As goldfish grew in popularity, breeders stepped up their game, creating ever more unusual varieties. Using artificial selection, they created goldfish with freakish and fantastical features, and the world's aquariums now contain the fantail, the veiltail, the butterfly tail, the lionhead, the goosehead, the golden helmet, the golden saddle, the bubble eye, the telescope eye, the seven stars, the stork's pearl, the pearlscale, the black moor, the panda moor, the celestial, and the comet goldfish, among others. This explosion of types was driven by the desire for the exotic and exquisite—urges that we can now satisfy with genetically modified pets.

We can also use genetic engineering to create animals that appeal to our aesthetic sensibilities, such as our preference for brightly colored creatures. For instance, a 2007 study revealed that we prefer penguin species that have a splash of yellow or red on their bodies to those that are simply black and white. We've bred canaries, which are naturally a dull yellow, to exhibit fifty different color patterns. And before GloFish were even a neon glint in Blake's eye, pet stores were selling "painted" fish that had been injected with simple fluorescent dyes. With fluorescence *genes*, we can make a true rainbow of bright and beautiful pets.*

Engineered pets also fit right into our era of personalization. We can have perfume, granola, and Nikes customized to our individual specifications—why not design our own pets? Consider the recent rise of designer dogs, which began with the Labradoodle, a cross between a Labrador retriever and a standard poodle. Though there's no telling when the first Lab found himself fancying the well-groomed poodle down the street, most accounts trace the

* Not all aesthetic alterations are created equal. Scientists have created beagles that turn ruby under ultraviolet light—by transferring a sea anemone gene into the dogs—but these GloDogs, as it were, are disturbing to gaze upon. They would surely be a harder sell than GloFish, perhaps because cough-syrup red is a color that never naturally occurs in the canine kingdom. Since nature itself has created some fish that are red and orange, however, artificially adding one of these hues to an aquarium resident doesn't seem so jarring.

origin of the modern Labradoodle to Wally Conron, the breeding director of the Royal Guide Dog Association of Australia. In the 1980s, Conron heard from a blind woman in Hawaii, who wanted a guide dog that wouldn't aggravate her husband's allergies. Conron's solution was to breed a Lab, a traditional seeing-eye dog, with a poodle, which has hypoallergenic hair. Other breeders followed Conron's lead, arranging their own mixed-breed marriages. The dogs were advertised as providing families with the best of both worlds—the playful eagerness of a Lab with the smarts and hypoallergenic coat of the poodle. The rest, as they say, is history. The streets are now chock-full of newfangled canine concoctions: puggles (a pug-beagle cross), dorgis (dachshund plus corgi), and cockapoos (a cocker spaniel–miniature poodle mix). There's even a mini Labradoodle for doodle lovers without lots of space.

Tweaking the genomes of our companions allows us to create a pet that fulfills virtually any desire—some practical, some decidedly not. When I set out to get a dog, I thought I had settled on the Cavalier King Charles spaniel: small, soft, and bred for companionship. Then I discovered a breeder who was crossing Cavaliers with miniature poodles, yielding the so-called Cavapoo. I was sold. I loved the scruffier, shaggier hair of the Cavapoo, and given what I knew about biology, I figured that a hybrid was less likely to inherit one of the diseases that plague perilously inbred canines. A dog that didn't shed would be an added bonus. Plus, poodles have a reputation for being brainy, and I'm an overachiever; if I was going to get a dog, I wanted to be damn sure he'd be the valedictorian of his puppy kindergarten class.

The hitch: Even the most careful selective breeding is a rough science. Sure, Labs are friendly and poodles are intelligent, but just letting them go at it doesn't guarantee that their puppies will exhibit the best of both breeds. Milo, the Cavapoo I eventually brought home, looks almost entirely like a spaniel, and as for a nonshedding coat, his health, and those famous poodle smarts? Well, my couch is covered with dog hair, Milo has a knee problem common in purebred Cavaliers, and I'm pretty sure he got the spaniel brain. So much for my plan to outsmart nature.

When I'm ready for my next pet, the landscape could be radically different. Social Technologies, a trend forecasting firm in Washington, D.C., issued a report on the commercial prospects for genetically modified pets. "Through advances in genetic modification," the report said, "biotechnology labs could join kennels and animal shelters as a source for the perfect pet . . . Initially a luxury, pet personalization would become available to the general public as the technologies involved become more mature."

Indeed, why bother creating clumsy crosses when we can edit genes directly? A company called Felix Pets, for example, is attempting to engineer cats that are missing a gene called *Fel d 1*, which codes for a protein that triggers human allergies.* And that's just the beginning. What if you could order up a fish created in your alma mater's trademark palette or dogs and cats with custom patterns on their coats? Or there's the ultimate designer pet, proposed by Alan Beck, director of Purdue's Center for the Human-Animal Bond: "If we're going to come up with genetically engineered animals, we might be able to come up with an animal that loves only you."

Transgenic pets will have to clear some hurdles before they make it to market. The Food and Drug Administration considers a new gene that is added to an organism to be a "drug," and regulates

* Another company, Lifestyle Pets, already sells what it claims are hypoallergenic cats. The cats, which go for nearly $7,000 a pop, are not products of direct genetic manipulation. Instead, the company says it has merely identified and bred cats with a natural mutation in *Fel d 1*. However, it remains unclear whether Lifestyle Pets has truly cracked the hypoallergenic code; controversy has long swirled around the company and its scientific claims.

altered animals under the Federal Food, Drug, and Cosmetic Act. Companies seeking approval to sell an engineered animal must demonstrate that the transgene has no ill effects on the animal itself. If the animal will be a source of food, companies must also demonstrate that it is safe for human consumption.

Regulators also evaluate how a genetically modified organism might affect the environment if it happened to make its way into the wild. Escape has been a concern since the first genetically engineered bacteria were created in the early 1970s. The scientists of that era worried about what might happen if they inadvertently created a dangerous superbug and it slipped out under the laboratory door. Biologists convened twice—at the Asilomar conferences of 1973 and 1975—to discuss these risks. In 1975, they drew up a document that encouraged their colleagues to exercise caution and use "biological and physical barriers" to ensure that novel organisms didn't break free from the lab. The National Institutes of Health issued guidelines stipulating such safeguards in 1976 and has periodically updated its recommendations over the years.

Though these containment strategies are now routine, they aren't foolproof, and ecologists continue to worry about engineered organisms ending up in the wild. Altered animals could "pollute" the gene pool by breeding with their free-range cousins, or snatch food and resources away from native organisms. In theory, laboratory manipulation could make a fish more likely to thrive in the big, wide world, and such Frankenfish could take over natural waterways, to the detriment of other species.

This very possibility has been part of the high-profile debate over the most famous (or infamous) transgenic fish: a fast-growing Atlantic salmon that AquaBounty, a Massachusetts firm, is trying to bring to market in the United States. Atlantic salmon normally produce growth hormone only in the summer, but the AquAdvantage fish have been engineered to crank out the hormone no matter what the season. The secret is a bit of biological code borrowed from the ocean pout, an eel-like fish that lives in frigid water. To keep its cellular machinery from icing over, the slithery fish produces its own antifreeze. The pout's antifreeze gene is normally attached to a sequence of regulatory DNA called a "promoter." Icy temperatures activate the promoter, which turns the gene on, triggering the ocean pout to start cranking out the antifreeze. The cold-sensitive promoter, however, can be attached to all sorts of different genes, and to create the AquAdvantage fish, scientists linked the promoter to a growth-hormone gene taken from the Chinook salmon. Then they slipped the entire construct into Atlantic salmon. As a result, in these salmon, cold temperatures prompt the production of growth hormone, and the fish reach their adult sizes faster than their unaltered counterparts. The genetic modification shaves a year and a half off the time between when a salmon hatches and when it's ready to garnish your bagel.

It's a clever bit of biological reprogramming, but AquaBounty has attracted vocal critics, many of whom fear that if the big bruisers from the lab escape, they could wreak havoc on wild salmon populations. To address these concerns—and reassure nervous regulators—AquaBounty is building several security measures into its production plans. It will breed fish in a secure facility in Canada and then raise the young in confined tanks situated in the highlands of Panama, far from their natural marine environment. The company also plans to produce only sterile female fish—incapable of passing their genes on even if they did somehow end up on the lam.

Though many scientists have concluded that there is little risk of the supersalmon escaping and staging some sort of wild coup, AquaBounty is still trying to win over regulators. The company first approached the FDA about its fish in 1993, and applied for formal approval in 1995. Despite deciding that the fish are low risk, the FDA has not yet ruled on whether they will be allowed on the market. (If the salmon are approved, they would become the first GM animal to officially enter the world's food supply.)

As Alan Blake prepared to bring GloFish to market, he studied the regulatory challenges that have hobbled AquaBounty. Blake wasn't sure what federal agencies would do about genetically modified *pets*, but he didn't want to take any chances, so he began calling government officials and asking whether they'd have concerns about GloFish. He told regulators that the fish were designed to be companions, not food, and reassured them that scientists believed the animals posed a negligible risk to the environment. Wild zebrafish, he told them, spend their time in the tropics, not the chilly waters of North America. Conventional zebrafish have been sold as pets in the United States for decades, and they have never been able to survive an aquarium jailbreak long enough to establish a wild population. The water is simply too cold, and the fluorescent varieties are even less likely to make a go of it—the data suggest that GloFish are more sensitive to cold temperatures, less successful at reproducing, and, one suspects, more visible to predators, with their big, neon EAT ME signs.

Of course, there is no such thing as zero risk, but Perry Hackett, a geneticist who studies zebrafish at the University of Minnesota, puts the danger posed by GloFish this way: "What are the odds that all the air molecules will rush up into a corner of the room you're sitting in and you'll suffocate? That for whatever reason, just at random, they all happen to collect just in one corner?" Such a scenario is theoretically possible, but it's so unlikely that we don't worry about it. As Hackett says, "We don't sit around with oxygen tanks by our desks."

Federal officials didn't register any serious objections in their conversations with Blake, and by the summer of 2003, he thought he had his bases covered. He had consulted with scientific and legal experts. The licenses to produce the neon Nemos were in hand. And the fish farmers were ready to start churning them out. Blake set a launch date of January 2004, but then California caught him by surprise. The state's Fish and Game Commission instituted a regulation prohibiting the production and sale of all genetically modified fish. Anyone who wanted to breed, buy, sell, or own these organisms needed to appear before the commission and request a formal exemption.

That fall, Blake was busy preparing for his hearing before the commission when a technical glitch suddenly made the company's password-protected website available to all eyes. The press got wind of Blake's Seussian fish, and within a week, the animals were discussed everywhere from National Public Radio to Al-Jazeera. Many publications ran anxiety-provoking stories, but the fearmongering award winner had to be a *New York Times* headline: WHEN FISH FLUORESCE, CAN TEENAGERS BE FAR BEHIND? As the story put it, "This is the tipping point, when the world irrevocably turns toward the science-fiction fantasies of writers . . . No doubt humans could be made to glow if parents with foresight knew that one day they would be desperately trying to find their middle school child at a dark and crowded school dance."

The stories made GloFish seem like monsters, harbingers of some sort of ethical or scientific apocalypse. Indeed, the genome can seem like a set of commandments—handed down and carved into stone—and fiddling with it makes us nervous. Selective breeding has become an accepted practice, but our ability to root around in the genome directly and move pieces of DNA between different species is still unsettling. "These are techniques that are advancing the threshold of human power over other species," says Richard Twine, a sociologist and bioethicist at Lancaster University. "It's a way of increasing the continuum of control over the animal and genotype and phenotype. There's an intensification, a new power that we didn't have before." What's more, once GloFish officially went on sale, they'd be available to anyone with five dollars, meaning that organisms once confined to pulpy science fiction novels could be living in your neighbor's den. With the launch of GloFish, biotechnology would come to our houses and knock on our front doors.

The California Fish and Game Commission seemed acutely aware of these concerns when it convened to discuss GloFish in December 2003. Unless you are an expert on the cold calculus of culling wild turkeys or an aficionado of the tender lovemaking habits of the New Zealand mud snail, Fish and Game meetings can be brain-deadening experiences. But on this particular afternoon, there would be a captivating showdown over our biotechnological future.

When Blake came to the podium for his opening remarks, he had a slightly bewildered air about him, like a straight-A student who suddenly finds himself called to the principal's office. He was well-mannered and deferential, peppering his comments with "sir"s and "gentlemen"s. As he spoke, it was obvious that he had done his homework. All the scientists that he had consulted—as well as the experts that the Department of Fish and Game had conferred with before the hearing—had concluded that GloFish were safe. But Blake had made a critical miscalculation: that the data would be enough.

GloFish may have been a laboratory triumph, but debates over biotechnology rarely come down to the science. According to public opinion polls, only 27 percent of Americans believe that the government should base its decisions about genetic engineering purely on science. Compare that with the 63 percent who think such decisions should take "moral and ethical factors" into account. That's just what the California commission did. Commissioner Sam Schuchat told Blake that he had already done a lot of thinking about whether GloFish should be sold in California. He'd even called his rabbi to discuss his concerns. "The question for me became an ethical question," Schuchat said at the hearing. "Here we are, playing around with the genetic basis of life, creating new organisms that don't exist. Now it is true that we human beings have been doing that for tens of thousands of years. But I guess at the end of the day, I don't think it's right to produce a new organism just to be a pet. I look at this issue in front of us and I think to myself, 'So, what's next? Pigs with wings? Pink horses?' "

"Let me be clear," he continued. "I'm not opposed to genetically modified organisms. But I don't think it is a good idea to employ this technology for what I would characterize as frivolous purposes . . . To me, this seems like an abuse of the power that we have over life, and I'm not prepared to go there today."

Blake had heard this objection before from some of the scientists he first consulted about his business plan. When Eric Hallerman, a fish geneticist at Virginia Tech, heard about GloFish, he worried that they were "a fairly trivial use of technology." But Hallerman, who has advised the federal government about risks that accompany genetically modified animals, overcame his initial skepticism, even joining the Yorktown Technologies Scientific Advisory Board. As Hallerman explains, when it comes to GloFish, "there's no harm being done, and there's fairly few enterprises that humans engage in, including agriculture, in which no harm is being done."

Let's not forget that even selective breeding can do harm. Those ornamental goldfish varieties that we've created to have eerie, unearthly eyes—enlarged and bulging, or covered by enormous growths, or positioned to look up toward the sky—can be nearly blind. From an ethical standpoint, isn't a fully functional transgenic fish preferable to an artificially selected but severely handicapped one?*

* We've also saddled dog breeds with all sorts of inherited diseases, and the English bulldog has been pushed so far by human selection that it is literally handicapped. The breed's massive head doesn't fit through the birth canal, and pups are usually born via cesarean section. Their snouts are so short that the dogs can barely breathe—they suffer from sleep apnea and a lifetime of oxygen deprivation. These breathing

Not, apparently, to the California commissioners. After they finished querying Blake, they voted, three to one, to deny his request. Commissioner Michael Flores was the lone dissenter. "We have a gentleman out here who's gone to the scientific community, those that are very precautionary, and they say that there's no risk," he said at the meeting. "So we're going to ignore that science, and that has me a little bit concerned." But Flores's single vote wasn't enough, and the objections of his colleagues meant that there would be no GloFish in the Golden State.

California could have been a huge market for Blake, who was disappointed with the ruling, but there were still forty-nine other states to sell in, and just days after the California commission rejected GloFish, the FDA released an official statement on the pets. It read, in full: "Because tropical aquarium fish are not used for food purposes, they pose no threat to the food supply. There is no evidence that these genetically engineered zebra danio fish pose any more threat to the environment than their unmodified counterparts which have long been widely sold in the United States. In the absence of a clear risk to the public health, the FDA finds no reason to regulate these particular fish."

A few opponents refused to accept the FDA's ruling as the final word. Just after GloFish hit pet stores in January 2004, the International Center for Technology Assessment and the Center for Food Safety—two affiliated nonprofits that have raised concerns about a variety of biotechnologies—filed a lawsuit. They alleged that the

---

difficulties also mean that the animals have trouble regulating their own body temperature, and many suffer early deaths from respiratory or heart failure. "If bulldogs were the products of genetic engineering, there would be protest demonstrations throughout the Western world, and rightly so," James Serpell, the director of the Center for the Interaction of Animals and Society at the University of Pennsylvania, once wrote. "But because they have been generated by anthropomorphic selection, their handicaps not only are overlooked but even, in some quarters, applauded."

FDA and the U.S. Department of Health and Human Services had shirked their legal duty to subject GloFish to a thorough review. In an attempt to convince the court that they had the right to sue, the plaintiffs constructed an unconventional argument. How had GloFish harmed them? Well, among other things, they said, the sale of the freaks of nature could lead to "aesthetic injury from viewing genetically engineered GloFish and other animals in aquaria . . ." The suit was eventually dismissed, but the "aesthetic injury" argument was a testament to just how desperate some opponents were to keep the animals out of pet shops. (Aesthetic injury? If that's a valid legal argument, I've got a couple of lawsuits I'd like to file. Mexican hairless dog, I'm looking at you.)

The aesthetic-injury argument apparently didn't find much traction with the public either, because GloFish, and their Kodak-worthy colors, are a hit, available in all of America's major pet store chains. (Yorktown Technologies sells its fish only in the United States, although the Taiwanese company Taikong sells its own version of the paint-box pets in Asia. Though he'd love to sell to customers in Canada and Europe, Blake doesn't want to tangle with these jurisdictions' ultratight restrictions on genetically modified organisms.) At first, Yorktown Technologies sold only red GloFish, but the company added green and orange varieties in 2006 and blue and purple in 2011. In 2012, the company introduced an entirely new fish: a white skirt tetra (*Gymnocorymbus ternetzi*) genetically modified to fluoresce bright green.* Petco, PetSmart, and Walmart also sell GloFish "kits," special tanks that come equipped with blue lights designed to bring out the fish's brilliance.

* Yorktown Technologies conducted "comprehensive" studies of the glowing tetra, Blake says, which revealed that the fluorescent tetra were less environmentally fit—and thus less likely to survive in the wild—than their unmodified counterparts. The company submitted this data to the FDA, which raised no objections to commercial sale of the fish, Blake says.

"We have e-mails from customers who love the fish," Blake tells me. "We've gotten thousands and thousands of e-mails and, on average, every year, we get—four? five?—e-mails from people that are expressing negativity. There are probably more people that claim to see Elvis flying a UFO in any major U.S. city every year."

Once GloFish hit the market, their fate was determined not by some abstract debate over biotechnology but rather by public demand. Customers simply like the fish. The success of GloFish is all the more remarkable in light of the public opinion surveys that show that most Americans aren't fans of lab-grown companions. (In one survey, 40 percent of respondents said that creating disease-resistant animals—such as chickens safe from the ravages of avian flu—was a "very good reason" to meddle with the genome. Compare that with the 4 percent who said that creating new pets was a "very good reason" to do so.) Is it possible that GloFish have changed our minds? Maybe there are some people out there who went into pet stores expecting something monstrous and came away thinking that GloFish were not only harmless but actually downright cool. It's what can happen when we get the opportunity to have close, personal encounters with biotechnology.

And it's one reason Blake takes his responsibilities seriously. Yes, he has a financial interest in GloFish's success, but he also knows that he has an opportunity to help shape public opinion. He hopes that GloFish will be a bright, shining example, proof that species engineering doesn't have to be so scary. "Biotechnology is often demonized," Blake says. "And then you see this tiny little fish, just swimming around, as happy as can be."

Are the fish happy? Are fish even capable of "happiness"? These are the questions I ponder as I stand, once again, at Petco, looking into a tank of glowing fish. It has occurred to me that just about the only thing I haven't done in the GloFish research department is invite them into my home. So here I am, ready to take the plunge. I grab the special GloFish aquarium and am about to pick out some plain gray stones to put in the bottom of the tank, but my boyfriend spots a bag of mixed gravel in hues that look like they belong on a tie-dyed T-shirt. "You should get those," he says.

"Won't that be tacky?"

"You're getting genetically modified, fluorescent fish," he says. "Don't you think that ship has sailed?"

I might as well go all in. I grab the fluorescent gravel and some neon plastic plants.

Then it's over to the corner tank that GloFish call home. They're swimming around in a hallucinogenic jumble, and I ask a clerk to corral six of them for me: two Electric Greens, two Starfire Reds, and two Sunburst Oranges. (At $5.99 each, I am stocking an aquarium with next-gen pets for less than $40—far less than the cost of my Cavapoo.) An employee plops the fish into a plastic bag filled with water. I hold the bag up to my face and come eye-to-eye with the doctored fish. They continue their openmouthed stares, hovering silently in the water. I don't exactly fear for the fate of the Earth. ("You'd think they were six feet long with fangs and they'd bite your head off, the way they've been portrayed," Blake once told me.)

I tote them home and set up the tank up in my living room. Under the blue light coming from the bulb, the GloFish gleam like jewels. I don't know if they're happy, but they certainly don't appear to be suffering. Neither am I—it's entrancing to watch them swimming around, a kaleidoscope in constant motion. These fish may be frivolous, but they're just a teaser, a preview of the coming attractions. If we can get black-and-white fish to glow neon red, green, and orange, what else can we get animal bodies to do?

2. Got Milk?

When scientists first learned how to edit the genomes of animals, they began to imagine all the ways they could use this new power. Creating brightly colored novelty pets was not a high priority. Instead, most researchers envisioned far more consequential applications, hoping to create genetically engineered animals that saved human lives. One enterprise is now delivering on this dream. Welcome to the world of "pharming," in which simple genetic tweaks turn animals into living pharmaceutical factories.

Many of the proteins that our cells crank out naturally make for good medicine. Our bodies' own enzymes, hormones, clotting factors, and antibodies are commonly used to treat cancer, diabetes, autoimmune diseases, and more. The trouble is that it's difficult and expensive to make these compounds on an industrial scale, and as a result, patients can face shortages of the medicines they need. Dairy animals, on the other hand, are expert protein producers, their udders swollen with milk. So the creation of the first transgenic mammals—first mice, then other species—in the 1980s gave scientists an idea: What if they put the gene for a human antibody or enzyme into a cow, goat, or sheep? If they put the gene in just the right place, under the control of the right molecular switch, maybe they could engineer animals that produced healing human proteins in their milk. Then doctors could collect medicine by the bucketful.

Throughout the 1980s and '90s, studies provided proof of principle, as scientists created transgenic mice, sheep, goats, pigs, cattle, and rabbits that did in fact make therapeutic compounds in their milk. At first, this work was merely gee-whiz, scientific geekery, lab-bound thought experiments come true. That all changed with ATryn, a drug produced by the Massachusetts firm GTC Biotherapeutics. ATryn is antithrombin, an anticoagulant that can be used to prevent life-threatening blood clots. The compound, made by our liver cells, plays a key role in keeping our bodies clot-free. It acts as a molecular bouncer, sidling up to clot-forming compounds and escorting them out of the bloodstream. But as many as 1 in 2,000 Americans are born with a genetic mutation that prevents them from making antithrombin. These patients are prone to clots, especially in their legs and lungs, and they are at elevated risk of suffering from fatal complications during surgery and childbirth. Supplemental antithrombin can reduce this risk, and GTC decided to try to manufacture the compound using genetically engineered goats.

To create its special herd of goats, GTC used microinjection, the same technique that produced GloFish and AquAdvantage salmon. The company's scientists took the gene for human antithrombin and injected it directly into fertilized goat eggs. Then they implanted the eggs in the wombs of female goats. When the kids were born, some of them proved to be transgenic, the human gene nestled safely in their cells. The researchers paired the antithrombin gene with a promoter (which, as you'll recall, is a sequence of DNA that controls gene activity) that is normally active in the goat's mammary glands during milk production. When the transgenic females lactated, the promoter turned the transgene on and the goats' udders filled with milk containing antithrombin. All that was left to do was to collect the milk, and extract and purify the protein. *Et voilà*—human medicine! And for GTC, liquid gold. ATryn hit the market in 2006, becoming the world's first transgenic animal drug.* Over the course of a year, the "milking parlors" on GTC's 300-acre farm in Massachusetts can collect more than a kilogram of medicine from a single animal. And so, the humble goat—tin-can eater and petting-zoo star—has added a new item to its résumé: pharmaceutical manufacturer. The universe of pharming is rapidly expanding; labs and companies around the world are working to stock their barns, fields, and coops with animals that pump out medicines for ailments ranging from hemophilia to cancer.† ATryn has now been joined by Ruconest, a drug produced in the milk of genetically engineered rabbits. Sold by the Dutch company Pharming, Ruconest treats hereditary angioedema, a genetic disease that causes painful bodily swelling.

Pharm animals, which push the boundaries of medical research and could save human lives, make GloFish look like child's play. There is nothing frivolous about them. But that's a double-edged

* The EU approved ATryn in 2006, and the United States followed suit in 2009. The gap in approval dates means that for three years, goats in Massachusetts were making medicine that could be prescribed only abroad.

† Scientists have also created genetically modified plants and bacteria that can produce some of these compounds. (In 1982, insulin produced by modified bacteria became the first genetically engineered drug approved by the FDA.) But many human proteins are complex—in order to work, they have to be folded correctly and adorned with special molecules—and animal cells are better than plants and bacteria at putting these finishing touches on a protein.

sword. Making animals more useful also makes them more likely to be *used.* Genetic engineering allows us to exploit other species for new reasons and in new ways, expanding our supply of creature commodities. Of course, using animals for our own purposes isn't new. Should we object because the technology is?

Scientists are working to coax all sorts of curative compounds out of animal bodies. Many of these substances are remedies for rare genetic disorders. James Murray and Elizabeth Maga—biologists and animal scientists at the University of California, Davis—on the other hand, have decided to use the tools of pharming to alleviate a much more pervasive problem: diarrhea. The ailment's global toll is enormous, with more than 2 million children dying of diarrheal disease every year. It's a ghastly statistic, and if Murray and Maga can begin to make a dent in that number, their work will be the most far-reaching pharming project yet.

As it happens, human breast milk is a potent antidiarrhea elixir. The liquid is full of compounds that boost a child's immune system and attack invading bacteria. The evidence now suggests that infants who are breast-fed have healthier digestive systems and are less likely to suffer from diarrheal diseases than those fed purely on formula. Some of these effects can last even after breast-feeding ends; infants who drink breast milk for the first thirteen weeks of life are less likely to come down with gastrointestinal problems during their entire first year.

One of the compounds responsible for these effects is an enzyme called lysozyme, a microbe destroyer that bursts bacterial cells like balloons, causing the cellular membranes to rupture and the disease-causing contents to spill out. Lysozyme is naturally present in the milk of all mammals, but it's especially concentrated in human breast milk, which contains three thousand times as much of the enzyme as the milk of some other animals. (Infant formula, which is usually made from cow's milk, has only trace amounts of lysozyme, at best.)

Murray and Maga want to extend the protective effects of breast milk to infants who don't nurse or children who have grown too old to do so. Their plan is to harness the power of pharming, engineering dairy goats that make extra lysozyme in their milk. The pair hopes that this genetically modified milk can be used to both prevent and treat childhood diarrhea. Like the scientists at GTC, Murray and Maga set out to create their supergoats using microinjection.* They squirted the human lysozyme gene into fertilized goat eggs and implanted the resulting embryos in surrogate mothers. One of the embryos grew into a little kid named Artemis, a transgenic female with a penchant for mulberry leaves.† She lives out at the university's goat barn, and one day, Murray took me to see her.

The barn is home to 150 assorted goats—representing a variety of wonderfully named breeds, including Alpine, Nubian, Toggenberg, and LaMancha goats—but Artemis has pride of place, making her home in a private enclosure directly in front of the entrance. Artemis, now a fully grown adult doe, is mostly white, with some black markings around her eyes and the classic goat accessory: a long white beard. As soon as we reach her pen, Artemis sticks her head in Murray's hands, waiting for him to stroke her ears. After Artemis matured, she became what's known as the "founder

* Microinjection has traditionally been the most commonly used technique for creating a transgenic animal, but it's not the only one. Scientists can also use modified viruses to infiltrate embryos and deliver transgenes. Alternately, they can insert a new gene into embryonic stem cells growing in the lab. These cells are then injected into an embryo; as the fetus grows, the adulterated stem cells will develop into tissues containing the new gene.

† In Greek mythology, Artemis is the goddess of hunting, wild animals, and childbirth.

female"—by breeding her, Murray and Maga generated a whole line of transgenic goats.

Today, Artemis's heirs are all over this facility, living in a line of wire-enclosed pens stretching out behind the barn. It's been raining all morning, and many of the goats are still huddled underneath their small wooden shelters. As we walk down the slick, hay-strewn path, the animals begin to "baa" and traipse over to us through the mud. I don't think I've had a close personal encounter with goats since I was a child, and I've forgotten how endearing the animals can be, with their wide-set eyes, their oversized ears, and their eagerness for attention and affection. The goats jostle one another, poking their noses through the holes in the wire fence, angling for pets from Murray and me. We happily oblige.

As we dole out our goat rubs, Murray points out the genetically modified animals. I'm glad he does, because they look just like all the other goats, and I never would have recognized them on my own. Eight of the transgenic females are pregnant, due to deliver a batch of new kids in the next month or two. With newborns to care for, these does' udders will fill with lysozyme-rich milk. They'll make up to two liters a day of the stuff, for about three hundred days after they give birth.

Murray and Maga have carefully analyzed the milk from Artemis's heirs and found that it does indeed contain elevated levels of lysozyme—1,000 percent more than normal. They've also demonstrated that the milk has protective effects in pigs, which have a digestive anatomy that is similar to our own. Compared with piglets that sucked down conventional goat milk, young pigs that got the special, transgenic stuff had lower baseline levels of coliform bacteria—including *E. coli*, a common cause of diarrheal disease—in their guts. They also had stronger immune systems and healthier small intestines. And when the researchers *tried* to make the pigs sick, by feeding them a delicious *E. coli*–laden soy broth, the piglets slurping down the lysozyme-rich milk fared better.

These results are giving Murray and Maga confidence that the modified milk will do a human body good. So in September 2011, they asked the FDA to review the milk from transgenic goats and officially rule whether it's safe for human consumption. They are still awaiting a verdict. Though Murray acknowledges that nothing can be 100 percent risk-free, he thinks that lysozyme is pretty safe. The compound is well studied and naturally present not only in our milk, but also in our tears and saliva. As Murray points out, "You've been eating lysozyme since the day you first swallowed."

Still, Murray and Maga aren't sure whether—or when—the FDA will rule in their favor. As of now, GloFish are the only transgenic animals available to the American public, and the federal government doesn't seem eager to give the neon swimmers any company. Ironically, it's the makers of useful transgenic animals—the ones designed to be sources of food or medicine—that are struggling to get the official stamp of approval. GloFish sailed through because they were totally trivial, designed purely to be pets. It is, of course, appropriate for an animal intended for human consumption to be subjected to a higher level of scrutiny, but the end result is that these organisms can get stalled in a never-ending regulatory process. Even if the FDA approves their transgenic goat milk, there's no guarantee that American doctors or patients will embrace it.

In the United States, the debate over genetic engineering has been dominated by loud antitechnology activists, and a stamp of government approval may not be enough to overcome pervasive fears about the safety—or ethics—of GM products. So Murray and Maga are hedging their bets and establishing a second herd of goats in Brazil, which is among a handful of countries—including Argentina, China, and India—poised to become major powers in the world of agricultural biotechnology. (The nation is already a leading grower of genetically modified crops.) The Brazilian government, which has been bullish on biotech, has given Murray and Maga's colleagues at the Federal University of Ceará $3.1 million to create their own herd of lysozyme goats. Once the goats are up and bleating, the international research team will start human trials in Brazil, studying the milk's effects in healthy adults and then healthy children. If all goes well, they'll move on to trials with those who might really be able to benefit from the milk.

The Brazilian team is headquartered in Fortaleza, along the country's northeastern coast. The region is home to some of Brazil's poorest towns and villages, where as many as 10 percent of kids die before their fifth birthday. Transgenic goats' milk may be a much-needed salve. Assuming the human trials are successful, Murray and Maga see the milk being used in several ways. Doctors could give it to infants who aren't breast-fed, in order to help them develop healthy immune systems. Or they could prescribe it for toddlers who are no longer nursing, in order to keep their guts in tip-top shape. Or the milk could be used as a *treatment* itself—administered alongside rehydration therapy to infants, toddlers, and kids suffering from diarrheal disease. As an added bonus, the milk would also provide some much-needed nourishment, combating the malnutrition that often goes hand in hand with intestinal disease.* The ultimate goal, Murray and Maga say, is to get their goats into towns and villages all over Brazil. Instead of keeping a herd of normal goats, families would raise transgenic ones, and anyone drinking the animals' milk would benefit from the supercharged levels of lysozyme.

* Murray and Maga haven't quite decided what the distribution system will be should the goat milk make it onto the market, though Maga says they probably won't sell the transgenic goat technology, or the rights to the milk, to a pharmaceutical company. Instead, she says, she and Murray have discussed partnering with a nonprofit organization to parcel out the milk.

Keeping children alive seems like such an unobjectionable endeavor, but doing so through genetic engineering makes many of us uncomfortable. The anxiety comes in many forms. There are legitimate concerns about health risks, but these are easy to address—that's exactly what human trials are for. The other objections are more philosophical—and harder to answer with cold, hard data.

Take, for instance, the worries about animal exploitation. The development of our new genetic tools has coincided with a growing concern for the rights and welfare of animals. In 1975, just as scientists were learning to mix and match DNA, Peter Singer published his famous treatise *Animal Liberation*. In it he railed against "speciesism," arguing that our mistreatment of animals, and our exploitation of their bodies for food or research, was akin to the subjugation of women or racial minorities. Animal suffering matters, he said, and we have an obligation to minimize the pain and distress we inflict upon other species. It was the birth of the modern animal rights movement, and in the years since, activists have launched a variety of diverse campaigns, lobbying governments to grant full legal rights to great apes and protesting companies that test cosmetics on rats or rabbits.

Those working to extend animal rights are motivated by a wide range of philosophies and goals, but one of the common themes is that animals have a basic *intrinsic* value—that is, that they are inherently valuable on their own, solely because they are living creatures with whom we share the Earth. On the other hand, when we use animals for food, or fiber, or drugs, we reduce them to their *instrumental* value, treating them as mere tools to be used or resources to be tapped.

Much to the chagrin of the animal rights crowd, biotechnology lets us turn animals into even better tools. With genetic engineering, we are giving animals the very traits we want to exploit. We are engineering lab rats guaranteed to suffer from the very medical afflictions we want to study and reprogramming dairy animals so they produce not only milk but also medicine.* As Richard Twine, the Lancaster University sociologist, notes of pharming, "Animals that have previously been defined as agricultural commodities are becoming pharmaceutical commodities. Biotechnology might be inciting new forms of commodification within human and animal relations. It's potentially multiplying the uses to which profits can be made from different forms of animal life."

For an extreme example, consider doctors' long-standing hope of using animals as organ donors for human patients. Worldwide, there's an acute shortage of human donors—in the United States alone, ten people awaiting new organs die every day—and animal organs could help fill the gap. Throughout the twentieth century, surgeons experimented with this kind of "xenotransplantation," putting monkey parts into humans suffering from various diseases and defects.† The most famous case came in 1984, when an infant

* In fact, these GM animals are beginning to take over scientific laboratories. Since 1995, the number of "genetically normal" animals used in British labs has decreased slightly, while the ranks of genetically engineered animals have grown more than sixfold. In 2010, 43 percent of all lab procedures performed in the United Kingdom involved GM animals. And Japan's labs are home to 3.6 million genetically modified mice. (While the USDA issues an annual report on the number of animals used in research, it does not break down animal usage by genetically modified status.)

† Not all such transplants were for life-threatening conditions. Take the popular procedure pioneered by the French surgeon Serge Voronoff. In the 1920s, he began performing an operation that he believed would keep aging men feeling young and vibrant. All an elderly fella had to do was get a thin slice of ape or monkey testicle sewed inside his scrotum. Thousands of men all over the world underwent the procedure, which was so popular that Voronoff worried about having enough animals to meet demand. (Learning that Voronoff also transplanted monkey ovaries into

with an underdeveloped heart received a replacement heart from a baboon. It was a daring experiment, but Baby Fae, as the patient became known, survived only twenty days with the new organ. Other recipients of monkey organs haven't fared much better. In the 1990s, for example, two patients received livers from baboons; one lived for seventy days after the operation, the other for just twenty-six. And none of the six patients who were given baboon kidneys survived longer than two months. The problem with these kinds of cross-species transplants is rejection; our immune systems expertly, and correctly, identify implanted animal parts as foreign tissues and attack the new organs.

Genetic engineering, however, may give us a way to create animal organs suitable for transplant into our own bodies. Using cognitively sophisticated monkeys and apes as organ donors has become taboo, so researchers are now focusing on pigs, which are widely farmed and have organs that are about the same size as our own. In fact, replacing defective human heart valves with pig valves has become a routine medical procedure. (This interspecies transfer of *tissue*, rather than a whole organ, is known as xenografting.) The surfaces of pig cells are adorned with signature sugars, which immediately tip our human immune systems off to the fact that something strange has entered the body. Surgeons can keep pig valves from provoking an immune response by treating them with a special preservative before they're implanted in the human body. It works fine for this small piece of tissue, but it's not feasible for entire organs, which must be fresh and viable when they're transplanted. Enter genetic engineering: scientists have now created pigs in which the gene that codes for these distinct sugars is

---

aging women is the first thing that has ever made me feel grateful to live in the age of Botox. Suddenly, injecting botulinum toxin into my face doesn't sound so bad.)

"knocked out," hoping that organs from these pigs will pass more easily as native human tissue.

Using these knockout pigs as organ donors could save thousands of human lives, but it also means that we'd be turning sentient creatures into more suitable sacrificial lambs, engineering animals purely so we can later dismantle them. That's instrumental in the extreme. Yes, we already break pigs into pieces to make our morning bacon, but genetic modification could expand the market for pig parts.

By and large, we accept the use of animals as objects and tools. Sixty-two percent of Americans surveyed in a Gallup poll, for example, deemed it "morally acceptable" to use animals for medical research, and despite the growth of the animal rights movement, there aren't many vegetarians.* And what does a T-bone steak represent if not a reduction of an animal to parts, to its instrumental value? There are issues with farming, of course, especially the industrial-scale factory farming that is the norm today. But whatever our objections to the system itself, the truth is that most of us accept the idea that we can use an animal's body to nourish our own.

For most of us, then, the real ethical question surrounding pharm animals comes down to the genetic engineering itself. Is there something about editing DNA and remixing biological material that is just inherently wrong? Monstrous mash-ups have long menaced the imagination, and critics of biotechnology worry that breaching species barriers violates the rules of God or nature or both. These concerns are magnified when researchers combine

* Polls reveal that between 97 and 99 percent of Americans at least occasionally eat animal flesh, and our appetite for meat is increasing—today, the average American eats a staggering 240 pounds of meat annually, up from 176 pounds in 1975.

animal DNA with our own, by, say, putting a human gene into a goat.

As it happens, some scientists are doing a whole lot more than inserting a single human gene into another species—they're creating human-animal "chimeras," whose bodies contain both human and animal cells. The difference between a transgenic animal, which has a single gene from a foreign species present in every cell, and a chimeric animal, which has cells that come from two different species, can be visualized this way: Imagine a transgenic animal as one in which every cell is blue with a single red dot, while a chimeric one looks more like a patchwork quilt, with some cells that are entirely blue and others that are entirely red. (To continue the analogy, a hybrid—created when the sperm of one species fertilizes the egg of another—would be a creature in which all cells are purple.)* For example, in a series of recent experiments, researchers at the University of Nevada, Reno, injected human stem cells—shape-shifting cells capable of becoming a variety of tissues—into sheep fetuses. As the lambs developed in utero, they incorporated these cells into their bodies, resulting in sheep that had hearts, livers, and pancreases that were part sheep and part human.

These interspecies combinations can raise uncomfortable existential questions, threatening our sense of uniqueness. If we can

* In the popular media, and public discussion, any creature that is part human and part animal is often referred to as a "hybrid," but technically, a human-animal hybrid is a very specific kind of creature: one created by fertilizing an animal egg with a human sperm (or vice versa). The most infamous attempt to create such a hybrid came courtesy of the Soviet scientist Ilya Ivanov. In 1927, Ivanov tried to impregnate female chimpanzees with human sperm, but when that didn't yield any tiny humanzees, he formulated a new strategy: He would inseminate Soviet women with sperm from Tarzan, a twenty-six-year-old orangutan. Happily for the women of the USSR, Ivanov was captured by the secret police before he could carry out this plan.

make our cells spring to life in a sheep or make a piece of our biological code work in a beady-eyed little rodent, what is it, exactly, that separates man from beast? Several states, including Louisiana and Arizona, have passed laws forbidding the creation of "human-animal hybrids," and U.S. Senator Sam Brownback has pushed for similar legislation on a national level. Brownback's proposed Human-Animal Hybrid Prohibition Act noted that "human dignity and the integrity of the human species are compromised by human-animal hybrids." (It's interesting to note that we rarely hear the flip side of this argument—that human-animal hybrids threaten the dignity of *animals*.)

From an ethical standpoint, human-animal mixtures are especially tricky when they involve the melding of minds. Animal cognition has much in common with our own, but certain kinds of autobiographical memory, language, number sense, and aspects of social cognition are unique to humans. At least, they are for the time being; scientists have already started manipulating genes involved in some of these capabilities. In 2009, German researchers engineered mice that carried a human version of FOXP2, a gene thought to be partially responsible for our unique way with words. (Mutations in the gene can cause speech and language disorders.) Giving mice the human FOXP2 variant changed the sound of the rodents' squeaks and the shape and size of their neurons.

What if, instead of making sheep with human cells in their livers, the scientists at the University of Nevada had made sheep, rats, or monkeys with a mass of human cells in their brains? Would these animals suddenly have a sense of justice? An ability to count? Would they be self-aware enough to realize that they were spending their lives as experimental subjects? If so, should we spring them from their cages? How many human brain cells and human behaviors would a sheep, rat, or ape need to display in order to qualify for enhanced legal status, legislative representation, and other rights?

Neither fully animal nor fully human, these creatures would occupy an ethical no-man's-land.

These sticky philosophical questions, among other concerns, led the Britain's Academy of Medical Sciences to conclude in its 2011 report that research that might make an animal's brain more "humanlike" should be subject to special scrutiny. Similarly, in the United States, the National Academy of Sciences has issued guidelines stipulating that any experiment that might cause human cells to end up in animal brains must have a strong scientific rationale to be approved.

Clearly, not all human-animal mixtures create the same quandaries. Putting a gene for human lysozyme in a goat does not make that animal a person any more than putting a pig valve into a human heart makes that surgical patient a pig.* Though both of these creatures are a mix of human and nonhuman animal, neither occupies some new, undefinable moral category. No one would seriously argue that a goat with a single human gene should get the right to vote or that a human with a pig part inside should be kept in a sty.†

Even as we worry about breaching species barriers, biologists argue over just what it is that makes a "species" in the first place. Though "species" exist as rigid categories in our minds—and are a

* That said, culture plays a role in how we view these human-animal combos. African scientists and policymakers have warned Murray and Maga that their transgenic goats may not be popular in certain African nations. In some cultures, the researchers were told, people would view the single lysozyme gene as enough to make the goats partly human and would consider consuming any part of those creatures to be a form of cannibalism.

† In the same report that recommended proceeding cautiously—or perhaps not at all—when it came to creating animals with human genetic material in their brains, the Academy of Medical Sciences concluded that "the great majority" of experiments involving animals engineered to carry human genes "pose no novel issues."

convenient way for us to label the natural world—they're considerably more fluid in nature. After all, Darwin's theory of evolution is based on the idea that there are smooth transitions, rather than sharp dividing lines, between humans and chimpanzees, between rats and rabbits. The genetic characteristics of a species are not set in stone; whatever it is that makes a human a human and a chimp a chimp is constantly evolving.

What's more, genes from different species sometimes mingle in the natural world. Animals occasionally pursue torrid interspecies affairs, giving us ligers and tigons and zorses. (Oh my!) Different species of bacteria can spontaneously swap DNA in the wild or transfer novel genes into insects, worms, and other animals. The parasite that causes Chagas' disease, a chronic illness associated with heart and digestive problems, can slip its DNA into our own genomes, and pea aphids have borrowed genes from a fungus that turns the bugs' bodies red. We can change animals faster and in more profound ways than nature does on its own, but the point is that there's nothing inherently sacred about a species's genome—it's an amorphous, ever-changing thing.

There is logic and then there's emotion. We don't have to believe that the genome is sacred or that humans are divine to feel a sense of revulsion when we imagine a mouse with a human brain. This reaction is what ethicists call the "yuck factor," and it's what makes us recoil when we consider the prospect of drinking wastewater (even after it's been decontaminated) or adopting a dog that glows neon red.

The bioethicist Leon Kass believes that we should pay careful attention to those visceral, gut responses to biotechnology. His essay "The Wisdom of Repugnance" was initially intended as a missive against human cloning, but his arguments have since been applied to all sorts of biotechnologies, including genetic engineering. "In critical cases," Kass wrote, "repugnance is the emotional expression of deep wisdom, beyond reason's power fully to articulate it . . . Repugnance, here as elsewhere, revolts against the excesses of human willfulness, warning us not to transgress what is unspeakably profound." Kass also argues that "repugnance may be the only voice left that speaks up to defend the central core of our humanity."

A knot in the throat or a pit in the stomach may suggest that we're approaching dangerous territory and need to consider our actions carefully, but we needn't let disgust run the show. After all, as an emotion, disgust is not always grounded in the world of reason. For example, the Academy of Medical Sciences discovered that while we're uneasy about giving animals human faces, limbs, hair, and skin, we're far less perturbed by making animals look human on the inside. This discrepancy, the report concluded, "appears to be irrational . . . [O]ne can compare this distaste at the humanised appearance of an animal with the common reaction of unease at the sight of human disfigurement. This is a primitive reaction which has no inherent 'wisdom.' "

Repugnance may be a good spark for public dialogue, but it shouldn't be a substitute for it. Acting in an ethical manner sometimes requires rising above raw emotion. What if we had let the visceral disgust some people once felt at seeing an interracial couple be the final word on interracial marriage? A gut instinct shouldn't be a death sentence, an emotional reaction a replacement for moral and ethical reasoning.

So if we discount the "yuck factor," how are we to evaluate the genetic alteration of animals? Bernard Rollin, a philosopher at Colorado State University, proposes that we use a simple ethic: "conservation of welfare." Simply put, he says, the principle holds: "If you're going to modify a line of animals, the resultant animals should be no worse off from a welfare point of view—and preferably better."

Some genetically engineered animals would certainly fail this test. The most infamous case is the "Beltsville pig," which carried the gene for human growth hormone. The goal was to create a pig that gained weight faster, required less food, and had less body fat than normal swine. The resulting transgenic pigs were indeed leaner and needed fewer calories to bulk up, but from a welfare point of view, the modification was catastrophic. The list of these little piggies' afflictions reads like a medical encyclopedia: joint disease, kidney disease, heart disease, diabetes, weakened immune systems, diarrhea, arthritis, ulcers, pneumonia, sexual dysfunction, and more. The swine also had bulging eyes and thickened skin, and they were lethargic and uncoordinated.

But not all genetic tinkering causes such animal welfare disasters. The precise effects of our engineering depend on the particular gene we insert and the snippet of regulatory DNA to which we attach it. In pharming, for instance, scientists have been able to restrict the production of foreign proteins to an animal's mammary gland by attaching it to a promoter active only in that part of the body. Our ability to limit a gene's activity to this one organ may explain why, by and large, pharm animals do not suffer from any unusual health problems. For example, the FDA examined seven generations of the ATryn goats, and found no evidence of strange ailments or illnesses. These goats have utterly normal lives—they just spend their days unknowingly secreting human medicine in their milk.

According to the conservation-of-welfare framework, the ATryn goats are ethically acceptable and the Beltsville pig is not. And the Beltsville pig is wrong not because it was genetically engineered but because it *suffered*. This ethical framework considers genetic engineering to be value-neutral—biotechnology is merely a tool, and whether it's a force for good or evil depends entirely on how we deploy it. As Rollin put it in his book *The Frankenstein Syndrome*, "It is simply false that all genetic engineering must harm animals. Unless one assumes that all species of animals exist currently at their maximal possible state of happiness or well-being of welfare, such a claim is not legitimate."

In fact, Murray and Maga's goats, which have not shown any signs of strange ailments or deformities, may be even *healthier* than their nontransgenic brethren. With higher concentrations of bacteria-busting lysozyme in their milk, the transgenic goats have healthier udders and fewer signs of infection, according to early data.

Other scientists are engineering livestock specifically to make them more resistant to disease. Several labs, for instance, have created cows that lack prions, the infectious proteins that can cause mad cow disease. In one approach, researchers used a technique called RNA interference. Messenger RNA, or mRNA, is essential for protein production—it carries the gene's instructions from the nucleus to the part of the cell where proteins are actually manufactured. Scientists have discovered that they can silence genes by injecting into a cell small molecules that destroy or disable mRNA while it's in transit. The RNA is thus not able to deliver its instructions to the cell's protein factories (it's as though the letter gets lost in the mail) and the protein is not produced. By designing molecules that target certain stretches of mRNA, scientists can silence specific genes and prevent the production of select proteins, such as prions. The prion-free cattle that result may be immune to mad cow disease.

Pharming continues to march forward. Biotech companies around the world are working on the next generation of transgenic dairy animals, capable of producing all sorts of crucial human antibodies, clotting factors, and other therapeutic proteins in their milk. Several teams of Chinese scientists have engineered cows that make milk with special nutritional properties, such as elevated levels of heart-healthy omega 3 fatty acids or reduced levels of hard-to-digest lactose. Researchers in some labs are working on producing transgenic animal drugs in other bodily fluids, such as blood, urine, and semen. (Apparently, a single boar ejaculation can contain a whopping 9 grams of protein. Talk about a "yuck factor.") A team of Japanese biologists got transgenic silkworms to spin cocoons containing human collagen. And a few researchers are putting their eggs in an entirely different basket: chickens. Thanks to our selective breeding for master layers, one hen can lay 330 eggs, each of which contains 3.5 grams of protein, in a single year. What if we gave these egg-making superstars jobs in the pharmaceutical industry? Scientists at Scotland's Roslin Institute have created feathered fowl that lay eggs containing compounds used to treat skin cancer and multiple sclerosis. Before long, we could all be cracking eggs for the cure.

The latest techniques are also making it possible to edit animal genomes with unprecedented precision. "The way we've been making transgenics up until now is really kind of crude," Alison Van Eenennaam, a geneticist at UC Davis, confesses. "You inject a bit of DNA and hope like hell it integrates somewhere on the genome. There are new technologies that are coming along that will allow us to go in and make very specific edits to the genome." One approach relies on what's known as "zinc finger nucleases"—lab-made proteins that act as little molecular scissors, cutting a strand of DNA at a specific location. Doing so means that researchers can disable one particular gene or slip a transgene into the genome in just the right spot. Today, scientists are far better equipped to control how a foreign gene is inserted and expressed than they were in the 1980s, which may help us engineer animals with fewer unwanted side effects.

Meanwhile, the emerging field of synthetic biology—in which scientists engineer new genes, cells, and biological systems from scratch—could eventually provide another way to design animals to our exact specifications. It's a young field, but it's moving fast; in 2010, the biologist J. Craig Venter announced that he had created a partially synthetic organism capable of replicating itself. Venter's team engineered the single-celled organism by building a genome that contained genes derived from a common species of bacteria as well as some entirely novel man-made stretches of DNA. (These custom-designed genetic sequences spelled out coded versions of the researchers' names as well as several famous quotations.) They inserted this genome into the cell of a different bacterial species, where it sprang into action, taking control of all cellular functions. Synthetic biology may yield new ways to build microbes—and, eventually, more complex life-forms—capable of producing drugs, biofuels, and other valuable compounds. (Of course, all the animal welfare, environmental contamination, and human safety concerns that accompany moving single genes around the animal kingdom are magnified a thousandfold when we consider the prospect of assembling an entire genome from scratch.)

Despite the scientific advances, political, economic, and social factors will keep some nations from embracing genetic engineering. European governments seem poised to reject products made by engineered animals, and the outlook in Canada and the United States is iffy. In 2012, Canadian researchers were forced to abandon fifteen years of research on an eco-friendly pig after their funding ran out. Researchers at the University of Guelph in Ontario had already engineered the animals, which they dubbed Enviropigs, to excrete less phosphorus, a common cause of water pollution. When phosphorus from animal manure makes its way into streams, lakes, and rivers, algae populations explode; this algal overgrowth can poison the water and kill fish and other aquatic organisms. Despite the pigs' potential, the scientific team was unable to find a company willing to bring them to market, and the animals were euthanized in May 2012. Animal rights activists had launched a campaign to save the pigs, and many people contacted the researchers offering to adopt the swine. But the scientists' hands were tied; regulations simply don't permit unapproved, experimental genetically modified animals to be released from a secure laboratory environment.

If other nations start approving, and possibly exporting, transgenic animal products, it will put pressure on the United States, Canada, and other nations to be more accepting of genetically engineered organisms. A careful review of transgenic animals is essential and can go a long way toward easing public anxiety, but it would be unfortunate if fears about genetic engineering, or arguments that the technology is inherently wrong, spurred governments to issue blanket moratoriums or let safe, useful animals languish in regulatory purgatory. That's what's happened to the AquAdvantage salmon, which the FDA has still not approved, despite concluding that the fish pose minimal danger to us or the environment. If the agency ultimately rejects the fish—or fails to approve them before AquaBounty runs out of money—it will have a chilling effect on biotechnological innovation in the United States, discouraging other scientists and entrepreneurs from developing new kinds of transgenic animals.

That would be a shame. Rejecting genetic engineering wholesale means that we'll lose the good along with the bad. And when it comes right down to it, as Murray says, "I don't think anybody in the world will turn down a drug from a transgenic animal if they need it or their loved ones need it. Or a transplant, if they need it."

It's easy to oppose biotechnology in the abstract, but when that technology can save your life, grand pronouncements about scientific evils tend to dissolve. Most of us would do a lot more than drink transgenic goat milk to have even one more day with our loved ones.

Or, in some cases, to spend more time with our beloved pets.

3. Double Trouble

Futuristic fantasies come in all shapes and sizes. Some of us may look forward to the day when we can bring home animals engineered to perform astonishing feats—a cat that glows while resting quietly on the couch or a cow that makes medicinal milk as it grazes in the backyard. For others, the dream might more closely resemble a hulking family man strolling into his local mall to pick up a replacement pet. That's the scenario that plays out in *The Sixth Day*, a sci-fi thriller set in the near future. In the movie, Arnold Schwarzenegger's character faces the sudden death of the family dog, Oliver. In response, he simply goes shopping, heading to a store called RePet, where a smarmy salesman offers to make Oliver's exact genetic duplicate. "Your RePet Oliver will be exactly the same dog," the salesman promises. "He'll know all the same tricks you taught him, he'll remember where all the bones are buried. He won't even know he's a clone."

In 2001, just a year after the film's release, a modified version of this sci-fi scene sprang to life when the world's first cloned housecat was born. The first cloned dog came four years later. Since then, animal lovers have welcomed Tabouli and Baba Ganoush (copies of a Bengal cat named Tahini), Lancelot Encore (the double of a yellow Lab named Lancelot), and a kennelful of other cloned kittens and pups into the world. The owners of these animals didn't want fanciful new animals—they simply wanted to re-create the old. It's an understandable impulse, familiar to anyone who's lost a cherished pet, and though just a handful of wealthy owners have had their animals cloned so far, as the science improves and the price drops, the market will grow.

If only bringing an animal back to life were as easy as that RePet salesman made it seem. Cloning is still a young, experimental technology—and one that raises serious animal welfare concerns. So before we order up our creature copies, we need to ask ourselves some thorny questions about what we can really expect from a DNA double and what costs we're willing to bear to get one.

We all know about the traditional, time-tested approach to babymaking: A sperm cell, bearing your father's DNA, meets an egg cell, which carries genetic code from your mother. When the sperm fertilizes the egg, the DNA mingles, and the embryo that results—the baby blastocyst that grows up to be you—is a biological cocktail. Half of the genes inside your cells can be traced back to Mommy dearest, while the other half come via dear old Dad. Cloning takes the normal rules of reproduction and turns them on their head; clones receive their entire biological inheritance from just one donor. Scientists can take a single cell from an animal's body—just the tiniest bit of skin, blood, or tissue—and use the DNA it contains to build a brand-new embryo. It's like taking the set of genetic instructions that gave rise to your mother, and plopping them, unchanged, into a fetus. A clone is essentially an identical twin born years after its genetic double.

The world of cloning changed forever on July 5, 1996, with the birth of a little lamb named Dolly. When Dolly was born, scientists had already cloned embryos, making exact genetic copies of unborn frogs, mice, and cows, but Dolly was revolutionary because she was the first clone of an adult mammal. Scientists at Scotland's Roslin Institute created her using a small tissue sample taken from a six-year-old ewe's mammary gland. The ewe had died years earlier, but the researchers just happened to have some of her preserved cells on hand, and they transferred the DNA from these cells into new sheep eggs. One of these eggs developed into a lamb that the researchers named Dolly (an homage to another fine mammary specimen: the country singer Dolly Parton).

Dolly was proof positive that researchers could take a small bit of flesh from a fully grown animal and create its identical twin, and her birth ushered in exciting possibilities in the reproductive sciences. Farmers and ranchers are constantly seeking to extend the genetic reach of their highest-performing animals, breeding like with like to create offspring that they hope will inherit the same milk-swollen udders or speedy legs. Cloning raises the prospect of making *exact* genetic copies of known performers, of creating perfect replicas of champion steers or horses that have proven their talents on the racetrack.

After Dolly's birth, scientists at Texas A&M University, in College Station, immediately recognized the implications. Like just about everything else, animal agriculture is bigger in Texas—the state has more cows than any other and the value of its livestock products ranks first in the nation—and A&M has an animal science department befitting this mammoth industry. The school has more than 700 acres dedicated to raising and researching cows, horses, sheep, and goats, and a dedicated Reproductive Sciences Laboratory, where scientists hone techniques—from artificial insemination to in vitro fertilization—that can help farmers manage the reproduction of their herds. When cloning came along, it gave the researchers a new tool for creating highly valuable livestock. In the years that followed Dolly's birth, the scientists at the lab proved cloning's potential, successfully carbon-copying a white-tailed deer, an Angus bull, a stallion, several litters of pigs, and more.

Along the way, A&M's researchers got involved in an endeavor they hadn't anticipated. Six months after Dolly's birth made international news, a man named Lou Hawthorne started recruiting reproductive scientists from America's laboratories. Hawthorne represented a wealthy client with an ambitious request: He wanted to clone a spayed dog named Missy, a border collie mix with a white face and silky gray coat. (Hawthorne's initially anonymous client was later revealed to be John Sperling, an eccentric billionaire who founded the for-profit University of Phoenix and has also bankrolled human longevity research. Missy belonged to Joan Hawthorne, Lou's mother and Sperling's longtime friend and lover.)

After considering a number of labs, Hawthorne picked a team of researchers at A&M for the dog duplication job. Mark Westhusin, a veterinary physiologist who ran the Reproductive Sciences Laboratory, would lead the cloning effort, and Sperling would fund the endeavor, to the tune of $3.7 million. When the Missyplicity Project was announced in 1998, pet owners flooded A&M with phone calls, asking about having their own dogs and cats cloned. It turned out that Sperling wasn't the only one who thought he had a special companion. As Hawthorne would later write: "Millions of people believe they have a one-in-a-million pet."

We no longer treat our pets as mere animals. We celebrate their birthdays and give them Christmas presents, let them lounge on the leather couch and sleep on the down comforter. Many of us consider our pets to be full-fledged members of the family, and their deaths prompt full-fledged outpourings of grief. We can seek the aid of pet bereavement counselors and choose from specialized caskets, headstones, and urns designed to send Fluffy off to the afterlife in style. So when word leaked out that researchers were trying to clone dogs, it naturally fueled hopes that we'd never have to let go of that one special friend, re-creating it—or at least its genetic double—over and over again.

The public reaction to the Missyplicity Project revealed a large potential market for copied pets, and Hawthorne and Sperling launched a company to make it rain cloned cats and dogs. On February 16, 2000, Genetic Savings & Clone (GSC) was officially born. At first, the company funded research and offered tissue banking, allowing people to store their pets' cells until cloning technology matured. (One page on the GSC website at the time suggested "a futuristic stocking stuffer: a gift certificate redeemable for the preservation of the animal's DNA . . . It's like a ticket to the future, today!") The company was an instant sensation.

Only one small thing lay between a pet owner and his clone: getting the darn ditto machines up and running. Although the impetus for the entire endeavor was a well-loved mutt named Missy, with both cat and dog owners clamoring for cloning, the A&M team decided to try replicating both species. Much to the disappointment of dog lovers everywhere, success with cats came first.

The lucky feline was a calico cat named Rainbow, and the first step in copying her was swiping a sample of her cells. When it comes to cloning, nearly any cell that contains a complete set of genes will do. (Recall that Dolly came from a mammary cell, and skin cells are also commonly used.) The A&M team knew that other scientists had had good luck with cumulus cells—the specialized, mature cells that surround a developing egg—so that's what they harvested from Rainbow.

But you can't simply stick a random cat cell into a uterus and expect a new feline to grow. Researchers needed to get Rainbow's biological code into the proper vehicle: an egg. To do so, the scientists employed a method known as somatic-cell nuclear transfer, the same approach the Scottish team had used to create Dolly. The technique involves extracting the DNA from an unfertilized egg and replacing it with instructions for making a clone. (The procedure is not unlike removing the custard from the middle of a Boston cream donut and refilling the donut with jelly.)

Westhusin and his team harvested ova from a clutter of lady housecats. They poked a tiny, turkey-baster-like tool called a pipette into each egg and sucked out its nucleus, leaving the rest of the cellular machinery untouched. The scientists took one of Rainbow's cells and placed it inside the newly "enucleated" egg, between the inner and outer membranes. This cell-inside-a-cell was then shocked with an electric current that turned the membranes of both cells into Swiss cheese, creating holes that allowed the genetic contents of Rainbow's cell to flow into the egg. The egg, thus "tricked" into believing it had just been fertilized by a sperm cell, began to grow and divide, just like a normal embryo.

The researchers ended up with three cloned embryos, each of which contained Rainbow's DNA. They transferred these embryos into the uterus of a brown housecat named Allie. Although only one of these feline fetuses survived to term, that was enough, and on December 22, 2001, Allie delivered a little, mewing kitten. Testing confirmed that the kitten was indeed Rainbow's clone, and she was given the name CC, short for "Carbon Copy."*

CC's name aside, technically, clones produced through nuclear transfer are not *quite* perfect copies of their genetic donors. Though the overwhelming majority of genes reside in a cell's nucleus,

* There's been some confusion about this name over the years, with many news outlets reporting—erroneously, according to Westhusin—that CC stands for "Copy Cat."

mitochondria—which produce energy for the cell and sit in the cytoplasm outside the nucleus—contain their own little genomes. Because nuclear transfer leaves the cytoplasm of an egg intact, CC had the mitochondria, and mitochondrial DNA, of her egg donor, rather than from her "twin," Rainbow. Since the amount of DNA involved is so tiny, however, most discussion of clones ignores this small genetic discrepancy.

CC's birth alone was a remarkable achievement, especially given cloning's staggeringly high failure rate. Some of the embryos created by nuclear transfer don't divide properly, some fail to nestle into the warm and welcoming uterine wall, some spontaneously abort themselves. In creating Dolly, for instance, the Roslin Institute researchers had made 277 attempts to make cloned embryos and ended up with just 29 viable ones. They were all transferred to surrogate mothers, and as time passed, the numbers dwindled further, until only one cloned fetus was left—the lamb that would be Dolly.

The challenges associated with cloning go beyond low success rates. Dolly died at age six, well short of a sheep's normal life expectancy. The lamb's creators maintain that her demise had nothing to do with cloning, pointing out that four other sheep in the barn also came down with the same contagious lung disease that took Dolly's life, but scientists are still plagued by questions about the health of cloned animals.

It's impossible to draw definitive conclusions about Dolly—or any other single case—but since her death, biotech companies have cloned hundreds of farm animals, and we've amassed much more data on the health of clones. The evidence is troubling. Failures and defects are a normal part of reproduction—not every fertilized egg implants in the uterine wall, and stillbirths and birth defects can happen no matter how an animal comes into being—and assisted reproductive technologies, such as in vitro fertilization, increase the risk of certain abnormalities. But clones, at least in some species, are more likely to suffer from birth defects and health problems than animals made by other means.

That's what the FDA concluded in a nearly thousand-page report, published in 2008, on the health of livestock clones. While the agency found no evidence of unusual health problems in cloned goats or pigs, it reported that cloned cattle and sheep do have an elevated risk of abnormalities. In particular, the animals are at risk for "large offspring syndrome," which can cause respiratory and organ problems in the newborns and complications for their surrogate moms. Cloned cows and sheep are more likely to die in the womb or shortly after birth than their conventionally conceived counterparts. However, the data reviewed by the FDA also showed that if the youngsters can be safely shepherded through the first six months of life, they seemed to develop into perfectly healthy adults, and when these clones reproduce the old-fashioned way, their offspring appeared to be normal. That said, the FDA also noted that "it is not possible to draw any conclusions regarding the longevity of livestock clones or possible long-term health consequences associated with cloning due to the relatively short time that the technology has existed."

Scientists believe that many of the poor outcomes seen in cloning can be traced back to a process known as genetic reprogramming. When a sperm cell fertilizes an egg, it initiates a cascade of changes. Some genes get turned on, while others are switched off, as the embryo grows and divides. Throughout the course of development, various genes are constantly being amplified or silenced, particularly as cells become specialized, or "differentiated." The activation or expression of certain genes equips a cell to join the heart, for example; the expression of different genes turns a cell into part of the skin, or the blood, or the brain instead.

For years, scientists thought cellular differentiation was irreversible—once a skin cell, always a skin cell. Dolly's birth smashed that assumption. Through the process of nuclear transfer, the scientists had managed to take the DNA from a differentiated mammary cell and turn it into something that a developing embryo could use. Cloning other adult mammals reinforced the discovery that nuclear transfer can reset genes contained in specialized cells back to their embryonic state. It was an astonishing accomplishment, turning back the genetic clock, but this process may not always go perfectly. As Westhusin explains, "An egg knows how to take a sperm cell, and the DNA that it has, and it knows how to reprogram it so that it turns some genes on and some genes off. A nucleus in a skin cell is not packaged like a nucleus in a sperm cell is. An egg knows how to reprogram a sperm to initiate life but it doesn't know exactly how to reprogram a nucleus from a skin cell."

Incomplete or flawed reprogramming can leave an egg with genes exhibiting an abnormal pattern of expression, which means that scientists might be creating a whole new cow with DNA that's stuck on the wrong settings. Depending on what genes are expressed abnormally, the result can be everything from an egg that's not even viable in the first place—and thus never develops into a fetus—to an array of birth defects. Though we don't know much about the health of cloned cats and dogs (there simply haven't been any large, long-term studies on the subject), errors in genetic reprogramming can affect any species.

Fortunately, CC was "vigorous at birth," all her little feline toes intact. For about a year, CC, Rainbow, and Allie lived at the lab, where the scientists monitored their health and showed off the trio to visitors. When the cats' scientific duties were complete, the researchers decided to place the felines in adoptive homes. Duane Kraemer, a veterinarian and physiologist who was part of the cat cloning team, claimed CC, and one December day, I went to meet her.

As I pull into a hotel parking lot in downtown College Station, I am nerdtastically excited. I'm about to meet my first clone! I pause to collect myself before heading inside to meet Kraemer; I want to play it cool. (Blurting out, "So let's go see the Frankencat!" would be a tad unprofessional.)

Kraemer is one of the university's senior scientists. He grew up on a dairy farm in Wisconsin and had planned to spend his life there, milking his family's cows. Then, as an undergraduate, he fell in love with research. He racked up five degrees—bachelor's degrees in animal husbandry and veterinary science, a master's in reproductive physiology, a PhD in reproductive physiology, and a degree in veterinary medicine—and joined A&M's faculty. He founded the Reproductive Sciences Laboratory and mentored dozens of students, including Westhusin. Now in his late seventies, he says he still gets a thrill every time he sees an embryo. He has a big smile, oversized glasses, and protruding ears, with a soft voice that sounds like it could give out at any moment.

We hop into Kraemer's car and my adventures in Cloneland begin. (Over the course of the day, I'll spend some quality time with CC, as well as snag a few minutes with Bruce—a cloned bull who would rocket a blob of his very valuable snot onto my sneaker—and Dewey, the world's first cloned white-tailed deer.) It's a short ride to Kraemer's ranch-style house. He ushers me around back, and as we push through the chain-link gate into the yard, his wife, Shirley, comes barreling out the door. I assume we'll be following her back inside the house to meet CC, but the couple points me in the opposite direction, to what looks like a big wooden shed in the yard.

"CC has her own house," Kraemer says. "For her and her kids and her husband."

Kraemer built the structure himself, and when he takes me inside, I am duly impressed. It's a bilevel, with a living room, a kitchen, and two small lofts. It has plumbing, heating, and air-conditioning. Should the cats ever get the urge to read, there's a set of bookshelves packed with the dissertations that Kraemer's students have written over the years. The back door opens into a screened-in patch of yard—cluttered with toys and branches—so the felines can get some sun and fresh air. It is, I have to admit, at least as nice as my own apartment. (What's a girl got to do to be reincarnated as a cloned cat?)

We find CC luxuriating on the landing. Kraemer walks over to give her a kindly rub, but she wriggles away and goes to sit on the windowsill, where she gazes out upon her kingdom. Her back sports gray stripes, and her belly, paws, and cheeks are bright white. She has green eyes and a small brown beauty spot, Cindy Crawford–style, just above her mouth. I stare into the face of pet cloning, and it stares back, twitching its soft pink nose. Despite not being a cat person, I have to admit that CC is—from a purely objective, scientific point of view, of course—pretty cute.

The Kraemers have given CC a good life—not only a house, but also a family. "We figured we should probably breed her because people would want to know whether clones could reproduce," Kraemer says. The matchmakers introduced CC to a gray tomcat named Smokey, and in 2006, CC gave birth to four kittens. One was stillborn, but the others were perfectly healthy.

As I wander around the cat house, I keep tripping over various members of the cat clan. One hangs out on a shelf in front of the AC, another claws furiously at a scratching post, while a third lounges languidly in a chair. CC keeps watch over the brood from her perch.

"I never thought I'd have a cloned cat," Shirley confides.

No? I laugh. That wasn't part of your life plan? "Is it strange?" I ask.

She pauses, then says, "Not as strange as when we had the lion."*

So far, CC shows no signs of health problems, cloning-related or otherwise, and a few months after I met her, she celebrated her tenth birthday. But there is something off about CC: She doesn't look like her genetic twin. Rainbow was a calico, her white fur splotched with gray and orange. CC, on the other hand, doesn't have a lick of orange in sight.

The most likely explanation for the discrepancy is a phenomenon known as "X inactivation." Like female humans, female cats have two X chromosomes. In calicos, the gene that codes for black fur is on one of these X chromosomes, while the orange gene is on the other. In any given cat cell, only one X chromosome is active. Westhusin and Kraemer surmise that in the cumulus cell used to create CC, the X chromosome carrying the orange gene was turned off.

CC is a reminder that DNA sequence isn't all that matters. An animal's characteristics also depend on how a gene is expressed. All along the genome, little molecular tags act as dimmer switches, turning genes on and off, making them more or less powerful. Some of these genetic settings are inherited and others are modulated by the environment. The chemicals and nutrients that a fetus encounters in utero, for instance, can make certain genes more or less active. Clones, carried to term by surrogate mothers, develop in different prenatal environments than their genetic donors did. Even after birth, early life experiences can alter gene expression in a multitude of ways. These environmental differences could easily lead to discrepancies between Fido One and Two.†

If you could get a clone like CC, who didn't even *look* like her

* Some years ago, Kraemer tells me, "a zoo brought a pair of lion cubs into the clinic and they only wanted one of them back." So, until A&M could build a suitable facility for her, Delilah the lion lived in the Kraemers' backyard.

† Errors in genetic reprogramming may also alter gene activity in a clone and account for differences between a clone and its genetic donor.

donor, you could certainly get one with a different personality. A&M's researchers saw that firsthand when they duplicated a Brahman bull named Chance. Chance was an unusually docile bull who'd starred in movies and on television, and his owner, a rodeo clown named Ralph Fisher, was desperate to have the bull replicated. In 1999, Westhusin made Fisher a clone, but as Second Chance grew up, it was clear that he was not the gentle giant his predecessor had been. Second Chance attacked Fisher. Twice. The second time, the bull pierced Fisher's left testicle, fractured his spine, and left the rodeo clown hospitalized with eighty stitches in his crotch. Second Chance had Chance's DNA in his cells, but he had a different upbringing, training, and life than his progenitor and became a different bull. (As for that RePet salesman's promise that a cloned pet will know all the same tricks as its DNA donor—pure hokum, at least in the real world.) CC and Second Chance are both illustrations of what became Westhusin and Kraemer's mantra: "Cloning is reproduction—it's not resurrection."

For his part, Kraemer was thrilled by CC's color snafu. He had been worried that pet owners might be easy prey for scammers. "People can be taken advantage of because of their devotion to their animal," he explains. CC was obvious, visible-to-the-naked-eye proof that even a genetic twin would not be a perfect replica of a special pet. But CC proved that cat duplication was possible, and in 2004, Genetic Savings & Clone launched its "Nine Lives Extravaganza," offering to mimeograph the cat of anyone who could afford the $50,000 price tag. The company also made a dramatic guarantee: "If you feel that your kitten doesn't sufficiently resemble the genetic donor, we'll refund your money in full with no questions asked." Less than a year later, the company delivered a cloned Maine coon cat named Little Nicky to its first paying customer. (The Texas woman who'd ordered the kitten was impressed. "He is identical," she told the press. "His personality is the same.")

Despite the success that GSC and A&M had with cat cloning, the company and the university eventually parted ways, in part because the A&M team encountered failure after failure in their attempts to duplicate Missy, the mutt that started it all. The vagaries of the canine reproductive system made the project more difficult than expected. In cats or cattle, immature eggs can be harvested from the ovaries and matured in a petri dish in the lab. For reasons that scientists still don't fully understand, this strategy doesn't work with dogs, whose eggs simply seem to be fussier. That means that researchers have to wait until the precise moment a canine ovulates, then open her up and flush the mature eggs out of her body. "The logistics of it are just a nightmare," Westhusin says. The A&M team managed to coax two canines to carry cloned embryos in their wombs, but one miscarried and the other gave birth to a stillborn pup.

GSC shut down in 2006 for financial reasons, but Hawthorne was soon back in business, at the helm of a startup called BioArts International. Still desperate for a Missy 2.0, he connected with Hwang Woo Suk, the South Korean scientist who had created the world's first cloned dog, in 2005, an Afghan hound named Snuppy (from *Seoul National University*, where the researchers were based, and *puppy*).* Hawthorne told Hwang, now at a company called the Sooam Biotech Research Foundation, about Missy and asked him to help give the dog a clone. Hwang, delivered—in triplicate—and by 2008, Hawthorne had three little balls of fur on his hands: Mira,

* Hwang has also been accused of fraud in connection with his claim, in 2004, of having cloned human embryos. (He was ultimately convicted of bioethical violations and embezzlement, but not of fraud.) Hwang reportedly admitted to falsifying data, and two of his landmark papers were retracted. His dog data, however, appears to be legit. (A spokesperson for BioArts defended the company's association with Hwang to *The Guardian:* "As a cloning company," he said, "we believe in second chances.")

Chingu, and Sarang, all clones of Missy.* The BioArts website announced, "Missy: Accomplished!" and noted that like their genetic donor, all three pups had soft coats and a fondness for broccoli.

Encouraged by this success, BioArts announced the "Best Friends Again" program, promising to have five dogs cloned by Sooam. The spots would be sold in a global auction, with bidding starting at $100,000. BioArts also announced a "Golden Clone Giveaway," in which a deserving dog owner, as decided by an essay contest, would win a free copy of his or her canine.†

However, the prospect of resurrecting pets didn't prompt universal excitement. Instead, it spurred the same kind of apocalyptic fantasizing that greeted GloFish. Animal replication sparked fears of copying humans, and a handful of journalists, ethicists, and politicians speculated about the potential for creating, say, an army of Hitlers. Some worried that cloning undermined individual uniqueness or that we were unleashing scientific powers beyond our understanding and control.

Other critics were concerned about animal welfare, an issue that deserves serious contemplation. Experts estimate that scientists worldwide use anywhere from 50 million to 100 million animals in their labs every year, and these creatures don't always have good

* Several experts have gone so far as to suggest that the secret to the South Koreans' success is the nation's appetite for dogs. Since the failure rates are high, a successful cloning attempt requires lots and lots of canine embryos. The Koreans have an advantage, some have said, because they have access to more dogs—and can harvest eggs from the canines being farmed or sold for their meat.

† The winner of the Golden Clone Giveaway was James Symington and his German shepherd Trakr, a search-and-rescue dog who worked the rubble of the World Trade Center in September 2001. Symington eventually received five clones of Trakr and founded Team Trakr, a nonprofit that will send teams of search-and-rescue dogs to assist in a variety of emergencies. All five of Trakr's genetic doubles are being trained to participate.

lives—think back to all those mutant mice, engineered to have cancer or Alzheimer's, or the hideous Beltsville pigs. Researchers sometimes inflict physical pain on their lab rats, performing invasive surgeries or exposing the animals to toxic substances. Lab animals can also suffer from psychological or emotional distress, caused by a lack of social contact and mental stimulation or forced participation in stressful experiments. As Marc Bekoff, a biologist at the University of Colorado, Boulder, who studies the inner lives of animals, explains: "Animals have the same desires that we have. They want to avoid pain, they want to just be content, they want to have their needs of food, shelter, friendship, sex, or whatever satisfied, and they want to avoid pain and discomfort and stress and terror. There can be no doubt about that."

Pet cloning tapped into long-standing worries about the burdens borne by lab animals, and the Humane Society of the United States and the American Anti-Vivisection Society joined forces to denounce the practice. In a 2008 report, the two groups sounded the usual alarm about the health of clones: "Few cloned animals are born alive, and many of those who do survive birth suffer health problems and die soon thereafter."

As the report pointed out, cloning's inefficiency raises additional welfare concerns. To duplicate one dog, scientists have to harvest eggs from a whole pack of anesthetized lady canines. Still more dogs are needed to carry the developing embryos in their wombs. (It takes a village. A furry, slobbery, tail-wagging village.) To create Snuppy, the South Korean researchers had implanted a total of 1,095 cloned embryos into 123 female dogs. Only two dogs were born, and only one lived. *Nature*, the journal that published the paper on Snuppy's birth, noted these sad statistics in an editorial: "It is unlikely that even the most obsessive pet owner would contemplate preparing more than 100 failed pregnancies for just one successful birth . . . In such circumstances, the cloning of dogs for pet owners remains ethically indefensible."*

The animals used in cloning, and other kinds of research, are afforded some protections. The federal Animal Welfare Act, which passed in 1966 and has been amended several times since, establishes basic requirements for the housing and care of laboratory creatures. It requires the use of painkillers and anesthesia when appropriate and stipulates that experimenters must consider both the physical and mental well-being of certain species. The law also includes special provisions for canines, social animals that thrive on human interaction. The act encourages researchers to provide dogs with "positive physical contact with humans," legally defined as "petting, stroking, or other touching, which is beneficial to the well-being of the animal." If the dogs are housed without any other canine compatriots, this extra human attention is mandatory.†

From the beginning, Genetic Savings & Clone made an effort to address welfare concerns by writing up its own strict code of ethics. Among other things, the code stated that all cats and dogs would get at least two hours of daily playtime, and, once their lab duties were over, that all would be placed in "loving homes." Any animal

* Hawthorne countered some of these concerns by saying that GSC would get its eggs by purchasing them from clinics spaying female cats and dogs, thus sparing healthy animals the burden of unnecessary surgery. The A&M researchers got most of their cat eggs from such clinics, but when it came to canines, the same approach "never worked out well," Westhusin says. "We never could figure out how to actually collect ovaries from a spay clinic and get these to mature in vitro to the point that they could be used for nuclear transfer."

† Universities and other scientific institutions are required to set up their own Institutional Animal Care and Use Committees to review research proposals and ensure that they meet the act's standards. Institutions that receive federal funds for animal research are also required to comply with additional welfare protocols, including those set out by the Institute for Laboratory Animal Research in its *Guide for the Care and Use of Laboratory Animals*, as well as the American Veterinary Medical Association's guidelines on euthanasia.

born with deformities would also be placed with adoptive families, unless the defects were severe enough to cause serious suffering, in which case the animals would be euthanized.

These promises weren't enough to calm critics, who insisted that any harm caused by pet cloning was unacceptable. After all, zebrafish and mice and goats are one thing—the thought of scientists experimenting with cats and dogs is much harder for us to stomach. Surveys show that more people disapprove of pet cloning than agricultural cloning, with about 80 percent of Americans opposed to duplicating pets in labs. (The percentage of those who say they disapprove of livestock cloning hovers in the mid-60s.)

Even the Animal Welfare Act reflects our preference for some species over others. While dogs are singled out for extra attention, many kinds of rats and mice—the very animals used in most experiments—are explicitly excluded from the act's protections. So are farm animals being used for "food and fiber." (Livestock being used for biomedical research are covered by the act, and the A&M team adhered to the federal law and additional welfare standards in its cloning work, according to Westhusin.)

Indeed, there's an interesting contradiction in the rhetoric that surrounds pet cloning. We're creating these carbon copies because we love our companions so much that we can't bear the thought of living without them. And yet that's also why the endeavor is so fraught—because we value cats and dogs above so many other species. People on both sides of the debate are driven by their devotion to these animals. The controversy over pet cloning is a debate over what it means to love an animal, and it involves values and judgments that we may never all agree upon.

Even the most stringent of ethical codes cannot guarantee that lab animals won't suffer; experimental procedures, by definition, have unknown outcomes, and cloning is clearly capable of causing animal pain and distress. Though cloners are figuring out how to improve the procedure's efficiency, they still have more to learn (particularly about the long-term health of lab-grown cats and dogs) before pet cloning is ready for prime time.*

Hawthorne ultimately came to the same conclusion, and on September 10, 2009, he announced that BioArts was getting out of the pet cloning business for good. In a statement that appeared on the company's website, Hawthorne wrote that animal duplication remained unpredictable. "Cloning," he acknowledged, "is still an experimental technology and consumers would be well-advised to proceed cautiously." What's more, he wrote, BioArts simply hadn't been able to attract enough customers. The company had sold just four of the five spots in its dog cloning auction. (Bio-.Arts successfully delivered clones to these four customers, plus the Golden Clone Giveaway winner, before shutting down its pet cloning operation.) Despite all the theoretical interest in copying a pet—the thousands of calls and e-mails—only a few owners were ready to pull the trigger.

Perhaps that's because the appeal of pet cloning is based an impossible dream, the fantasy put forth in *The Sixth Day*—the hope that through the miracle of science, we can bring a beloved pet back to life. Compare this motive with the cold, hard calculus behind agricultural cloning, which is not about love, but money. A cattle rancher simply wants to create twins of animals with superlative musculature or milk production. That's an achievable goal. Several cloned cows, for instance, have won the World Dairy Expo, the largest dairy show in the United States. Doc, the winning steer at the 2010 Iowa State Fair, was a clone of the animal that won the same competition in 2008. It's not only cheaper and easier to clone

* One small piece of evidence for improvements in efficiency came in 2011, when news broke that Dolly had been cloned again. Four genetic copies of the infamous sheep are alive and well in Scotland. While it required twenty-nine cloned embryos to create Dolly, each of these four new clones required just five embryos.

a bull than a dog—$20,000 for the cow, compared with $100,000 or more for the pooch—but it's also a better investment; a genetically gifted bull can bring in so much money that the cloning more than pays for itself.* (What's more, when cloning came along, ranchers were also used to thinking about breeding in scientific terms and had grown accustomed to managing their herds with the help of reproductive technology, and the world of livestock breeding was already home to companies eager to commercialize the latest laboratory breakthroughs.)

There's enough demand for cloned livestock that ViaGen, a company based in Austin, Texas, is cloning several hundred farm animals a year.† Most of its customers are duplicating cattle, but horses are poised to be the next big thing. ViaGen has cloned a champion barrel racer, and an Argentinian polo player has had some of his high-performing horses copied. And in 2012, the Fédération Equestre Internationale, the governing body for international equestrian competitions, reversed its ban on cloned horses, clearing the way for such equines to compete in the Olympics. (Imagine a race in which all the horses are clones of a previous winner. Or clones of the same winner! What a challenge that would be for the odds makers.)

A pet owner who pursues cloning is motivated not by the desire to monetize a single physical trait, but by a devotion to a unique

* Despite the furor over cloned meat, actual *clones*—which cost a lot to produce and are stuffed with great genes—are unlikely to become hamburger. They're simply too valuable to their owners to be slaughtered. And although the FDA concluded that meat from clones is unlikely to pose any elevated food-consumption risks, the USDA has issued a voluntary moratorium, asking owners of cloned livestock to keep the animals out of the food supply. Instead, cloned cows will be used primarily as breeding stock, and their offspring, conceived in the normal way, will end up on our dinner plates. In fact, since the FDA does not require milk or meat from the offspring of clones to bear any special labels, such products may already be in grocery stores.

† One of ViaGen's financial backers? Our good friend John Sperling.

animal with a whole suite of characteristics and quirks. Though there's more to producing a prizewinning cow than genetics, cloning is simply better suited to fulfill the more limited goals of farmers than the grand dreams of pet lovers. Even a biological double will never bring a pet back from the Great Kennel in the Sky. That makes it hard to justify paying six figures for a clone, especially while the technology is experimental and the results unpredictable.

The dream is still alive, however, and pet lovers haven't given up hope of seeing old Rover again. What if GSC and BioArts failed because they were simply ahead of their time? Over the coming decades, it's likely that cloning's success rates will go up, the price will come down, and public attitudes will soften.* Several animal gene-banking companies are betting on it; for a small fee, these companies are offering the same service that GSC did in its earliest days, letting pet owners put their animal's DNA on ice until cloning technology matures.

One of these companies, PerPETuate, was founded all the way back in 1998, and it's still going strong. The company's website makes the hard sell, assuring prospective customers that it provides the opportunity to "produce extraordinary, matchless, one of a kind, physically superior, brilliant, and innately talented replacements for lost pets." I called up the co-founder and president, Ron Gillespie, to find out how it all works. Gillespie guided me through the process. If I wanted to store my Cavapoo Milo's cells, he told me, the company would send me a tissue collection kit. With the help of my vet, I'd take "two small punches" of skin from the scruff of Milo's

* A 2011 Gallup poll shows that respondents between the ages of eighteen and thirty-four are more likely than their elders to view cloning as morally acceptable, a trend that may drive further acceptance.

neck and mail the samples back to PerPETuate. In the company's lab, technicians would isolate the skin cells, let them reproduce like crazy, and then tuck them away in a tank of liquid nitrogen. Milo's cells would hibernate in this stainless steel "Bio-Kennel," sitting in a deep freeze alongside DNA from other pet dogs, cats, birds, and lizards.

PerPETuate plans to offer its clients the opportunity to transform these cellsicles into clones of their furry friends when the technology becomes more reliable and less expensive. In the meantime, several of its customers have had their dogs' cells sent to South Korea, the de facto capital of canine cloning. Sooam, BioArts's former partner, continues to crank out dog doubles, as does RNL Bio, another South Korean cloning company.* But few people can afford to shell out six figures for a Korean clone, and Gillespie says the price will have to drop considerably (to $10,000 or less) before he adds cloning to PerPETuate's list of services.

Still, animal lovers keep sending in DNA samples, eagerly awaiting the day when they'll be able to order a clone. When we spoke, in fact, Gillespie had just gotten a call from a woman in Florida who had a rat—he emphasized this word when he told me the story: "a *rat*," he said—that she wanted to preserve. Sadly, the rodent, dead by the time its owner had called PerPETuate, no longer had viable cells. The owner insisted the tissue be stored anyway, Gillespie said, "because it gives her hope." As PerPETuate's website reassures potential customers: "There is almost no fee that would be too much to ensure the possibility of replacing your beloved pet with a twin sometime in the future." After all, you can't put a price on love. (But

* RNL Bio also has plans to expand—a recent press release made cryptic reference to a planned "theme park for cloned dogs." I assume this involves creating a place where we can interact with cloned canines, but I prefer to imagine a carnival where the duplicated dogs themselves can unwind, riding Ferris wheels and chowing down on funnel cakes.

if you had to, it would be somewhere in the neighborhood of $1,300, plus a yearly storage fee. All major credit cards accepted.)

Despite the visions of mini-Milos scampering around in my head, I don't think I'll be sending my dog's cells to PerPETuate anytime soon. Cloning simply wouldn't give me another dog exactly like Milo, and even if it could, I wouldn't want one. When Milo's gone, I'll want to start over with an unrelated dog, free of expectations and unencumbered by constant comparisons between the old dog and the new.

But if we understand cloning's limitations—and researchers figure out how to create healthy, thriving clones with less collateral damage—I don't begrudge pet owners the right to make their own choices. We all have different values when it comes to caring for animals, and our bonds with our pets are full of emotion. Must grieving dog owners have a logical reason for wanting Fido's DNA to live on? Making pets in a lab is not strictly "necessary," but when our shelters are jammed full of unwanted canines, most dog breeding is unnecessary, too. Is one method of creating new animals really more abhorrent than the other?

I hope cloning outcomes improve, because we have more to gain from the technology than a few doubles of dead pets. Westhusin, for instance, cloned a bull that had natural resistance to brucellosis, a common cattle disease. Lurking somewhere out there in the wider world may be cows that are resistant to mad cow disease or chickens that are immune to avian influenza. Cloning these genetic freaks could lead to populations of farm animals that are healthier themselves—and safer for us. We might be able to use the same approach to create healthier pets. Imagine starting a new breeding population of Labrador retrievers with clones of dogs that are free of the hip problems that often plague these canines.

Then there are the wildlife biologists who have been racking their brains for ways to boost the populations of endangered species. Zoos have been running breeding programs for decades, but captivity makes it difficult for many animals to get in the mood. The work is arduous and the results inconsistent. So wildlife breeding specialists have taken note of the reproductive technology that's revolutionizing livestock breeding. They have paid close attention as scientists learned to make DNA doubles of sheep, cows, cats, and dogs. And they have decided to borrow the technique to help threatened critters make a comeback.

4. Nine Lives

It's tough to be a wild animal on Earth today. The world is teeming with 7 billion humans, and our wants and needs—for housing developments, cheap food, and the latest and greatest electronic gadgets—are destroying what's left of the wilderness. Almost a quarter of the mammalian species roaming the planet are at risk of extinction; the same goes for almost one in three amphibian species and one in eight species of birds. There have been five mass extinctions in history—the most recent is the one that wiped out the dinosaurs—and many scientists believe we're at the beginning of a sixth. Conservationists have been doing what they can to preserve habitat, but it's like trying to bail out a boat that's constantly sprouting new holes; demographers estimate that, by 2050, there will be more than 9 billion people on Earth.

So it's not surprising that scientists have started searching for alternatives, looking to biotechnology for potential solutions to the extinction crisis. A few iconoclastic researchers think they've found one in cloning. On the surface, the idea is simple: *Animal numbers dwindling? Let's just use science to make copies of the ones that remain!* But it will not be nearly as easy as it sounds. That much has been apparent since the birth of the very first endangered-species clone: a little gaur named Noah, an exact copy of a rare wild ox native to India and Southeast Asia. His birth, in January 2001, was a headline-grabbing feat, proving that it was at least technically possible to mimeograph endangered animals. It was also a bittersweet accomplishment. Thirty-six hours after he was born, Noah began showing signs of a gastrointestinal infection. Twelve hours later, he was dead. The researchers at Advanced Cell Technology, the Massachusetts company that brought Noah into being, said cloning had nothing to do with the calf's tragic fate, but it's impossible to say for sure, given the health problems that have been documented in other clones. Noah's death suggested that wildlife replication would not be immune to the challenges and complications that have plagued those cloning pets and livestock. But when it comes to endangered species, there's a pretty compelling rationale for forging ahead. Cloning these rare animals is about more than money or companionship—it's about survival.

With these high stakes in mind, I decide to take a trip to New Orleans, where a small group of researchers have positioned themselves at the forefront of endangered-species cloning. Their remarkable facility is hidden inside 1,200 acres of hardwood forest along the banks of the Mississippi River. At first glance, these woods look like any other slice of nature. But peek inside these thickets and you'll find some surprising secrets: Some of the world's most exotic animals—creatures that usually make their homes on the African savanna or in the mountains of Central Asia—are here, living quietly in this small patch of wilderness. Amble among these trees and you could find yourself face-to-face with a flock of preening snow-white ibises or a small spotted wildcat, pacing back and forth.

These are the grounds of the Freeport-McMoRan Audubon Species Survival Center. The entire compound sits at the end of a country lane, behind a locked gate. A guard checks my credentials, then allows me in. I drive slowly along a narrow gravel road that winds into the forest. Branches swoop over me, creating a lush canopy, and it's impossible to see more than a foot or so past the trees that line the road. I half expect a leopard to leap out in front of my car at any moment.*

Suddenly, the forest opens up into a clearing, where a brick sign welcomes me to the sprawling Audubon Center for Research of Endangered Species (ACRES), the Survival Center's 36,000-square-foot complex of genetic and veterinary laboratories. Each of the rooms inside is devoted to one small task in the much larger effort to save wild animals. Signs posted on the doors along one corridor announce, in succession: GAMETE/EMBRYO LABORATORY, MOLECULAR GENETICS LABORATORY, RADIO ISOTOPE LABORATORY, CRYOBIOLOGY ROOM. For a state-of-the-art research facility, it feels awfully homey, with its dark wood paneling and bucolic views. I have just settled down into a plush armchair when Betsy Dresser, the reproductive physiologist who directs ACRES, emerges from her office. Wearing a gray blazer the same color as her closely cropped hair, she offers a warm handshake and a smile.

Dresser has spent her life in the company of other species. As a child, she constantly begged her family to take her to the nearby Cincinnati Zoo, and as soon as she was old enough, she began working there, moving up through the ranks from teen guide to zoo-

* I would later discover that there is, indeed, a clouded leopard hidden in these woods, though it's tucked away safely inside a cage.

keeper to junior zoologist. When Dresser went to college in the 1970s, she discovered the field of reproductive biology. She read the latest work coming out of Duane Kraemer's lab at Texas A&M, in fact, and watched as scientists learned to manage herds of cattle through careful breeding, artificial insemination, and other reproductive technologies. But while scientists were futzing about with farm animals, populations of the world's wild creatures were beginning to plummet. As Dresser recalls, "I saw the science and technology coming forward for our domestic animals, and I kept thinking, 'Why can't we do this with wildlife? Why can't we apply some of this to at least try to save some species?' "

After earning a PhD in animal reproductive physiology, Dresser established the Center for Conservation and Research of Endangered Wildlife (CREW) at the Cincinnati Zoo in 1981. At CREW, Dresser and her colleagues made a number of breakthroughs, including producing a Persian leopard cub through artificial insemination and creating the world's first test-tube gorilla. Impressed by the research at CREW, the Audubon Nature Institute, which ran a zoo in New Orleans, asked for Dresser's help in creating a similar program. In 1996, Dresser found herself leading the brand-new Audubon Center for Research of Endangered Species (ACRES). "We exist because we want to see wildlife in the future," Dresser says of ACRES. "I can't imagine just seeing elephants and lions and tigers just in textbooks, like we see dinosaurs today."

To make sure that these species stick around, Dresser, who directed ACRES for fifteen years and continues to consult with its scientific team, is willing to use whatever reproductive technologies are at her disposal. At first, the ACRES crew relied on the same techniques Dresser had honed in Cincinnati—embryo transfer, in vitro fertilization, and the like—and the walls of the research facility are hung with photos of tiny kittens and newborn whooping cranes that the scientists brought into being. Dresser acts every bit the proud parent, showing off each creature. "Here's a caracal," she says, pointing to an image of two kittens she created using in vitro fertilization. The otherwise sand-colored cats have tufts of black fur jutting straight out of the tips of their pointy ears. "Some people," Dresser says, "call them Spock cats."

Dresser goes down the line of photos, identifying each kitten: serval, fishing cat, Arabian sand cat, and more. Nearly all of these felines are under threat, a result of poaching and habitat destruction. The ballooning human population is hurting these cats in other ways, too—our pet tabbies and Persians can't seem to keep their furry little paws off their undomesticated cousins. This freewheeling interbreeding creates litters of cute little hybrids, but it doesn't help boost the ranks of wild felines.

ACRES has made a name for itself for its work with these small exotic cats, and as the technology evolved, so did the scientists' strategy. Though in vitro fertilization had allowed biologists to help exotic animals breed in new ways, the technique had limitations. Creating test-tube caracals, for instance, required harvesting sperm and eggs from wild cats, fertilizing the eggs in the lab, and then implanting them in surrogate mothers. Collecting and storing specialized reproductive cells is technically difficult—and potentially dangerous for the animals, since the females must be anesthetized and cut open in order for surgeons to recover their eggs.

Cloning has several distinct advantages. Scientists can get all the DNA they need for cloning from an animal's skin cells; stealing a quick swipe of skin from a rare cat is a much easier proposition than surgically harvesting ova. Cloning also provides a way to propagate the genes of animals without viable sperm or eggs: old animals, infertile animals, even dead animals. To Dresser, the technique has an obvious role to play in rescuing endangered species. As she imagines it, scientists could collect skin samples from rare animals and then churn out new copies of them in the lab. Field biologists could take these creatures and release them into their native habitats, where the clones would mingle with their wild brethren—both socially and sexually—and the population would slowly rise again.

The prospect of using cloning to save endangered species is a big dream, one that will require many researchers and many years to pull off. So Dresser and her colleagues are beginning with the basics. They're not running large-scale repopulation projects. They're not even trying to produce hundreds of endangered clones. Instead, their role is to nail down the technology itself—to test cloning in different species, fine-tune the laboratory procedures, and publish the results. That way, Dresser says, "If a habitat can't be saved, and a population isn't breeding naturally, and numbers dwindle to where there're only five individuals left in a species, we can call on these tools."

For all her optimism, Dresser is also a realist: She knows that cloning alone will not be enough to save a species. The ACRES team isn't, for instance, tackling the environmental problems that are causing the extinction crisis in the first place, but Dresser believes that reproductive technology is an important piece of the puzzle. As she argues, "There is no one answer to saving endangered species or wildlife on this planet. There are many very good organizations in this world that work on saving habitat. And that's what they do best. Why not do what *we* do best? My passion burns where I think we can be part of the solution—emphasis on *part* of the solution."

For her first cloning project, Dresser chose the African wildcat (*Felis silvestris lybica*), a tawny-colored feline with black rings circling its legs and tail. Native to northern and western Africa, the animals are thought to be the ancestors of domestic cats. Dresser decided to duplicate a three-year-old African wildcat named Jazz who already resided at ACRES, and technicians began by taking a tiny sample of the cat's skin cells. To do the cloning, the researchers planned to employ nuclear transfer—the same technique that researchers had used to create Dolly, CC, and others—but with a twist. Usually, scientists put the DNA of the animal being duplicated into an egg harvested from a female of the same species. When the scientists at A&M cloned Rainbow, for instance, they put her genes into an empty egg taken from another domestic cat.

Nuclear transfer presents extra hurdles for wildlife biologists, who may not be able to get their hands on enough females of an exotic species to provide eggs or act as surrogate mothers. And even if they rounded up a pack of wildcats, they'd be loath to put threatened animals through any unnecessary medical procedures. So when scientists clone endangered animals, they usually use a common, closely related species to serve as egg donors and surrogate mothers. This is known as *interspecies* nuclear transfer.

To clone Jazz, Dresser and her colleagues used everyday housecats. They collected ova from plain ol' tabbies, removed the nuclei, and then used the standard nuclear transfer procedure to put Jazz's genes inside. The eggs from the domestic cats now contained instructions for building a wild one.* To maximize their chances of

* Animals created through nuclear transfer, you may recall, aren't quite perfect replicas of their DNA donors, because they contain the mitochondrial DNA from their egg donors. And so, using interspecies nuclear transfer to duplicate an endangered species raises an interesting philosophical question. As the Rutgers University biologist David Ehrenfeld put it in a 2006 essay, "[I]s a cloned animal, whose mitochondrial DNA is at least partly from the egg donor species, a true copy of the species we are trying to conserve; and does it matter if it is . . . ?" It's a provocative question, but in the long run, scientists could keep the foreign DNA from spreading through a wild population with a little careful breeding. Since mitochondrial DNA is inherited entirely from the mother, all researchers would have to do is prevent the female offspring of female clones from having kittens. Male offspring of female clones, and all offspring of male clones, could reproduce freely.

success, the researchers implanted the cloned embryos in fifty different lady housecats, and twelve ended up pregnant. The ACRES team carefully monitored the pregnancies, using regular sonograms to check on the developing kittens. Alas, cloning's inefficiency reared its ugly head, and it was a long and sometimes heartbreaking slog. The first three cats miscarried. One went into premature labor; the kitten did not survive. Several kittens were stillborn. A few more survived their birth, but died within thirty-six hours.

The string of losses was eerily similar to what other cloners had faced, and the incomplete genetic reprogramming associated with nuclear transfer likely contributed to these poor outcomes. But the ACRES team kept at it, and on August 6, 2003, they extracted a tiny wildcat kitten—weighing less than a stick of butter—from the womb of a housecat named Brooke. The vet cleared the male kitten's nose and mouth and watched him take his first breaths. As soon as Brooke was sewn up, the staff placed the kitten beside her, and the newborn started to nurse. The researchers watched and waited, hoping that when the anesthesia wore off and Brooke came to, she'd bond with the fuzzy ball of foreign DNA pressed up against her.

The odd couple thrived. Brooke took to her maternal duties like a champ, and the little clone continued to suck down her milk. After several uneventful days, Dresser and her colleagues let out a sigh of relief; it looked like the youngster would make it. In a nod to their New Orleans home, they named the kitten Ditteaux (pronounced, in the French fashion, as "Ditto"), and DNA analysis confirmed that he was, indeed, an exact genetic replica of Jazz.

Ditteaux soon had company. That November, Miles and Otis, two more clones of Jazz, were born, as was Caty, a copy of a female African wildcat named Nancy. Spring brought four more Nancy duplicates: Madge, Emily, Evangeline, and Tilly. All the clones were raised by their surrogate mothers, and when they reached sexual maturity, they became swingers, mating in various combinations: Ditteaux and Madge, Ditteaux and Nancy, clone with clone. Their kittens were normal and healthy, and many were eventually sent to live at various zoos.

After these successes, the ACRES researchers moved on to other small exotic cats, cloning the caracal and the Arabian sand cat, the same species that Dresser so proudly showed off when we first met. Next up: lions and the Canadian lynx. They've created the cloned embryos already. All that's left to do is implant them in surrogate mothers. Meanwhile, other labs and researchers have been busy making their own breakthroughs. A European team made a mouflon, a rare breed of wild sheep, using DNA extracted from a female found dead in a pasture, and Korean researchers cloned an endangered cattle species as well as the gray wolf. In 2012, scientists in India ushered Noori, a clone of the rare pashmina goat, into the world.

But that doesn't mean we're ready to stock the wild with clones. For every well-earned accomplishment, there are disappointing setbacks; nuclear transfer still produces failures and casualties, whether scientists are duplicating pets, livestock, or wildlife. After making Noah, Advanced Cell Technologies went on to clone the banteng, another variety of endangered cow from Southeast Asia. The first such banteng, born to a domestic cow, was perfectly healthy, but its identical twin, born to a different cow two days later, was hugely oversized at birth. It was a classic case of the "large offspring syndrome" that can plague cloned calves, and the second banteng was euthanized when it was a few days old. If we want to use clones to prop up a population, we'll need to figure out how to produce healthy animals with less collateral damage and learn more about the long-term health of clones. (Ditteaux is still alive and well at age eight, and as scientists rack up more successes, and more clones come of age, we'll have the chance to close this knowledge gap.)

If and when we are ready to use clones for large-scale repopulation projects, what would such an endeavor look like? How would we go from a cloned kitten living in a lab to a sustainable wildcat population? In Dresser's mind, the first task would be simply to create a lot of wildcats. Biologists would collect skin samples from as many of the felines as possible and send them off to a facility like ACRES. The laboratory scientists would turn the skin cells into cloned embryos, and a few months of gestation would turn the embryos into wide-eyed wildcat kittens. But researchers couldn't just let the clones loose; repopulation projects are major undertakings, requiring long-term scientific, economic, and political commitments. Captive-born wildcats would need to learn survival skills, such as how to hunt on their own, and biologists would need to work with African governments and agencies to secure a safe slice of land for the felines. That wouldn't be an easy task given that it's habitat destruction and other forms of human interference that have gotten small exotic cats into trouble in the first place, and the clones might need to start their wild lives on a sanctuary or preserve. After the cats were released, scientists would need to spend years monitoring the animals, analyzing mortalities and documenting how the lab-born cats were adjusting to their new lives. If all went well, the cloned felines would eventually integrate themselves into the wild population and begin to breed.

Many endangered-species reintroductions fail—reviews have turned up success rates that range from 11 to 53 percent—but there have been some important victories. Such programs have boosted the wild populations of black-footed ferrets in the United States, golden lion tamarins in Brazil, and Arabian oryx in Oman, among others.

In addition, animal reintroductions can have ripple effects that help restore the environment itself. Every species is part of a complex ecosystem, and if an animal population suddenly disappears—or its numbers drop precipitously—it can throw the entire system out of whack. For example, some plants rely on animals to disperse their seeds; if these animals die out, the plants are vulnerable, too. When large herbivores disappear, dry shrubs and grasses accumulate, increasing the chance of wildfires. When predators disappear, herds of grazing animals swell, stripping the landscape of vegetation. Some scientists have proposed that by reintroducing animals to their native habitats, we can remodel landscapes and restore healthy ecosystems.

One researcher is putting this idea into action in the northern tundra of Siberia. Today, it's a desolate place, the snow-covered ground featuring little vegetation beyond shrubs and moss. But it wasn't always this way. During the Pleistocene epoch, which ended some twelve thousand years ago, the tundra was thick with wild grasses. Woolly mammoths, bison, and wild horses roamed the land. According to Sergey Zimov, the director of Russia's Northeast Science Station, these large herbivores played a key role in maintaining these grasslands. "In the winter, the animals ate the grasses that grew the previous summer," Zimov wrote in *Science*. "All the while they fueled plant productivity by fertilizing the soil with their manure, and they trampled down moss and shrubs, preventing these plants from gaining a foothold. It is my contention that the northern grasslands would have remained viable . . . had the great herds of Pleistocene animals remained in place to maintain the landscape."

Zimov is trying to turn back the clock by bringing the Pleistocene's major herbivores—or their modern equivalents—back to the tundra. The animals will be deposited in Pleistocene Park, a large preserve that Zimov established in northern Siberia. Zimov hopes that these large herbivores will help convert the moss-covered landscape back into grassland and restore the diversity of plants and animals that have since disappeared from the region. The project will unfold over the course of decades, but reindeer, moose, musk oxen, bison, and wild horses are already wandering the park and beginning to shape the landscape.

There are more radical proposals, too, such as "rewilding" North America by stocking the Great Plains with wild horses, camels, elephants, cheetahs, and more. (Elephants would serve as stand-ins for mammoths and African cheetahs as proxies for the extinct American cheetah.) According to the scientists championing the idea, these exotic animals will help convert weedy, rat-infested landscapes into lush, biologically diverse grasslands. (And, one imagines, turn a simple trip to Walmart into a drive-by safari.)

The ultimate effects of these ambitious projects are unknown, but even a small-scale reintroduction can help restore an ecosystem. Take the once-abundant gray wolf, which had disappeared from Yellowstone National Park by the mid-1920s. The park's elk populations exploded in the decades that followed, and these hungry ungulates spent their days chowing down on aspen, willow, and cottonwood trees, stripping branches of their leaves and chewing through saplings.

In 1995 and 1996, officials took a few dozen wolves from Canada and released them into the park. The wolf population slowly grew, and the elk population shrank back to a sustainable level. Today, the vegetation is recovering as well: the trees are taller, and the leaf canopies are thicker. This, in turn, has made the area more hospitable to other species. Songbirds are more abundant, and beavers, which had all but vanished from the park, are returning. What started out as a modest reintroduction project has been restoring Yellowstone to a place where all sorts of species can thrive together.

In the long run, boosting population size is just part of the task for scientists such as Dresser, since many endangered species are also handicapped by a lack of genetic diversity. Consider the enormous variation among human beings, all the different traits possessed by the people in your family, or in your state, or in Mozambique, Sri Lanka, and Iceland. Imagine that a meteor hits Earth and spares (miraculously!) only the people living on your block. Whole family trees, and their unique genetic variants, have been wiped out. You and your neighbors are the only people who can repopulate the planet, and even if you reproduce in every possible combination, your descendants will not be as genetically diverse as the human race was before the meteor.

This reduced diversity creates all sorts of problems. It means a rare and devastating mutation might proliferate. If, purely by chance, the genetic mutation that causes Huntington's disease is hiding in your neighbor's genome, your block's descendants could suffer from the disorder in staggering numbers. And if there are only a few families contributing their DNA to the communal pool, there will inevitably be a lot of inbreeding, which can cause problems of its own. A small gene pool also invites other disasters; if an infectious disease comes roaring along, and everyone's equally susceptible to it, it could wipe out humanity in one fell swoop.

This is essentially what happens to a species whose numbers have dropped precipitously, such as the cheetah. Evidence suggests that some unknown catastrophe wiped out most of the planet's cheetahs about ten thousand years ago, leaving just a small number of the cats to pass their genes along. The cheetahs alive today are a remarkably homogeneous bunch, with very little genetic variation. Their low levels of fertility and high rates of sperm abnormalities may be a result of generations of inbreeding.

Cloning, which just makes twins of the creatures that are already out there, won't solve the genetic-diversity problem for cheetahs or any other species, but we could use the technology to prevent a gene pool from shrinking further. For instance, if scientists learn to clone the cheetah—something they have not yet attempted—they could create carbon copies of the animals that don't reproduce. If one of the wild felines dies in infancy, and scientists can get their hands on a skin sample, they could clone the youngster, giving it another chance to pass along its genes. They could do the same with cheetahs that reach old age without ever having little ones. In a small population, every genome counts.

By stockpiling DNA from exotic animals, we can also prevent other species from developing such crippling diversity problems in the first place. At ACRES, these DNA samples are kept behind the door labeled "Cryobiology Room," and Dresser takes me inside. The room is cold, dark, and unimpressive. There is no obvious high-tech lab equipment, just a cluster of metal tanks, the approximate size and shape of kegs, lined up along the wall. But appearances can be deceiving. "That is years of science in there," Dresser says, gesturing at the tanks.

This is the Frozen Zoo, where an entire wild kingdom is packed into a few square feet. Dresser opens one of the tanks, which are kept at a frigid –373 degrees Fahrenheit, and nitrogen vapor comes swirling out. Suspended in the fog is a metal rack packed with tiny yellow straws. Each straw contains a cell sample from a different animal. They hold skin cells, sperm, eggs, and whole embryos from thousands of different individuals, including gorillas, elephants, rhinos, monkeys, buffalo, frogs, storks, cranes, lions, tigers, and bears. If Dresser is a modern-day Noah, these tanks are her ark. (Indeed, the story of Noah pops up again and again in the world of endangered-species cloning, from the little cloned gaur that was named after the biblical figure to the researchers who invoke the tale when discussing their DNA banking projects.) When the samples are frozen just so—pumped full of a cryoprotectant that prevents cells from bursting as the temperature drops—they can survive indefinitely.

Frozen zoos provide us with the opportunity to preserve the genetic diversity of a species *before* catastrophe strikes. If they had existed when the cheetah population was at its most robust, scientists could have packed the tanks with hundreds or thousands of cheetah skin samples. If we had these cells available today, we could look through them for genetic variants that have disappeared from the wild. We could clone these animals back into existence, and set them free on the African savanna, restoring genetic lineages that had died out.

According to Duane Kraemer—who established his own project to bank wildlife DNA after his students showed an interest in species conservation—what we really need to do is store cells from animals that are *not* endangered, preserving their diverse DNA for the future. "We should be systematically sampling populations and putting their cells into storage," he says. "Our species has a tendency to wait until we're in trouble until we look for solutions."

We can change that with frozen zoos, which are popping up all over the planet. The San Diego Zoo has a particularly well known one, and eighteen institutions in eight countries are participating in the Frozen Ark Project, run out of Britain's University of Nottingham. Together, these institutions have collected and preserved 48,000 DNA samples from more than 5,500 species; the collective goal is to hit 10,000 species by 2015. If we store these samples properly, we'll be able to use them to pull off remarkable scientific feats, including resurrecting species that die out in the wild. For example, the San Diego Zoo's ice-cold collection includes cells from the po'ouli, a small Hawaiian songbird believed to be extinct. (The last known po'ouli died in 2004.) Scientists haven't yet figured out how to clone birds, but if they do, the po'ouli DNA is sitting there in liquid nitrogen, ready to be brought back to life—and to give the fabled phoenix a run for its money. (Who needs a mythological bird that rises from the ashes when we've got a real one?)

The closest that scientists have gotten to species resurrection is the cloning of the Pyrenean ibex, a Spanish mountain goat. By 1999, there was only one Pyrenean ibex left in the world. Her name was Celia, and the rest of her kind had been hunted to extinction. One day in January 2000, Celia found herself under the wrong tree in Spain's Ordesa National Park. The tree toppled, crushing Celia and officially snuffing out the Pyrenean ibex for good—or so it seemed.

The year before Celia's death, some forward-thinking researchers had swiped a sample of her skin and stored the cells in liquid nitrogen. Then, after the elderly goat was gone, the scientists thawed out her cells and used nuclear transfer to get the ibex DNA into a whole clutch of domestic goat eggs. For their surrogate mothers, the researchers used hybrids—female crosses between the domestic goat and the Spanish ibex, a subspecies closely related to the Pyrenean variety. After the five-and-a-half-month gestation period, one hybrid was still pregnant. Researchers opened her up and delivered Celia's clone. The newborn kid opened her eyes and moved her legs, but she struggled mightily for air, and died just a few minutes after her birth. A necropsy revealed lung abnormalities, a defect that's been observed in other young clones. It was the briefest of resurrections, but the return of the Pyrenean ibex gave scientists hope that cloning could indeed bring back other extinct species.*

Several labs have embarked upon projects to clone species that died out long before Celia. Australian researchers want to revive the thylacine, one of the many strange creatures that evolved in the land down under. Like a kangaroo, the thylacine was a marsupial that carried its young in its pouch, but it looked more like a hyena, and the dark brown stripes running down its back earned the animal one of its other names: the Tasmanian tiger. (It's also sometimes called the Tasmanian wolf.) The mammal has been extinct since 1936, when the last thylacine died at the Hobart Zoo. The technology used to painstakingly preserve cells in frozen zoos was not around when the thylacine disappeared, but we've held on to a few strange souvenirs: dried thylacine skins and wrinkled, hairless thylacine pups floating in alcohol-filled jars.

Clearly, these are not ideal conditions for DNA, which degrades over time, but a few Australian scientists think that they can use these samples to clone the Tasmanian tiger. They haven't done so yet, but other researchers have managed to get some decent DNA from these thylacine samples. In 2008, one team of scientists isolated a piece of DNA from a baby thylacine that had been stored in alcohol a hundred years ago. Then they put the thylacine fragment, which controlled bone and cartilage formation in the animal, into the genomes of mice. The DNA jumped right

* One dream is to resurrect Lonesome George, the famous Galapagos tortoise who died in 2012. George was the planet's last Pinta giant tortoise, and after his sudden death, scientists hustled to preserve some of his cells. The president of Ecuador said he hopes researchers will clone George, but before that becomes possible, scientists will need to learn much more about the reproductive biology of tortoises, as well as figure out how to clone reptiles.

back into action, performing its normal regulatory duties in the bodies of these transgenic mice. Triumphant, the scientists wrote: "[W]e have restored to life the genetic potential of a fragment of this extinct mammalian genome." The following year, a different group of researchers got their hands on some thylacine hair and published the complete sequence of two Tasmanian tigers' mitochondrial genomes. These were exciting developments for those who dream of seeing packs of striped marsupials hunting down wombats and wallabies. But let's not count our Tasmanian tigers before they hatch; the deterioration of the DNA in our thylacine samples means that bringing the animal back to life is still a long shot.

The longer a species has been extinct, the more difficult the resurrection. That makes the oft-cited goal of cloning a woolly mammoth—last seen alive circa ten thousand years ago—an especially daunting one. In recent years, several mummified specimens have been discovered under the Siberian permafrost. The ice has helped preserve the carcasses and, scientists hope, the DNA they contain. Russian, Japanese, and Korean researchers—including the (in)famous cloner Hwang Woo Suk—have joined forces to extract DNA from these carcasses and re-create the prehistoric giants using elephants as egg donors and surrogates. (During his 2006 trial for fraud, embezzlement, and other alleged offenses, Woo Suk admitted that he'd spent some of his research funding trying to buy mammoth tissue from the Russian mafia.)

The scientists have a . . . well . . . mammoth task ahead of them. To use nuclear transfer, they'll have to find a cell in pristine condition. That will be a difficult task; the passage of thousands of years, cycles of freezing and thawing, and the presence of various microbes can all damage genetic material, and the DNA in even the best mammoth specimens unearthed so far has shown evidence of degradation. The other option—to sequence enough different genetic fragments to yield a complete, error-free genome and then build a set of chromosomes from scratch—is even more daunting. Add to that all the normal challenges that accompany cloning, plus the difficulties of working with the elephant reproductive system. (Among other obstacles, researchers will need to navigate more than *eight feet* of reproductive tract to get a cloned embryo inside an elephant's uterus.)

If that doesn't sound tough enough, we could try going back even further, to the Jurassic era. Dinosaur DNA is way too far gone for cloning, but the famed paleontologist Jack Horner has a different proposal for bringing back the reptilian beasts. Birds, scientists now know, are the modern descendants of dinosaurs. In fact, the genomes of birds and dinosaurs are so similar that Horner thinks we can reverse-engineer the reptilian beasts from chicken embryos. To make a "chickenosaur" that looks like a prehistoric raptor, we wouldn't even have to add new genes to a chick embryo—we'd just have to alter how its current genes were expressed, Horner says. Put a bird cell in a dish and bathe it in just the right growth factors, and we might be able to run evolution in *reverse*, prodding chicken DNA to build something that looks like it belongs in *Jurassic Park*.

Unfortunately, even if we can surmount all the technical issues, bringing extinct animals back might prove to be more cruel than kind. What would become of a resurrected mammoth or a Tasmanian tiger—or even two or three? They would be mere curiosities, carnival creatures confined to labs and zoos. Life in the wild likely wouldn't be much better. Though Zimov has kindly offered up Pleistocene Park as a potential refuge for any future mammoth clones, we'd be sending the animals out into a world vastly different from the one they once knew. We might be setting the animals up for a miserable existence on a planet that can no longer give them what they need.

Cloning concerns environmentalists for just this reason—because the technology allows us to just churn out new animals without restoring or repairing their habitats. To many biologists, cloning is all sizzle and no substance, a high-tech spectacle that fails to address habitat loss, poaching, pollution, and the other human activities that put wildlife at risk in the first place. David Ehrenfeld, a biologist at Rutgers, raised this concern in an article in *Conservation Biology.* Cloning, he wrote, "is a glamorous technology, and there is the danger of creating the false impression in the mind of a technology-infatuated public that it offers an easy, high-tech solution to the problem of extinction. Not only can this divert resources from conservation methods that have a much better chance of success, but repeated cloning failures may disillusion the lay supporters of conservation." Cloning, he concluded, "should never be a conservation strategy of first resort."

But the time for first resorts has come and gone, and safeguarding species is an all-hands-on-deck enterprise. Indeed, for cloning to have a real shot, laboratory scientists must work with conservationists; researchers can make all the fauna facsimiles they want, but the lab babies will need somewhere to live. The prospect of unleashing a thousand clones in the planet's forests and prairies may be pure fantasy, but it's not so far-fetched to imagine using cloning to accomplish more modest goals, such as duplicating select animals from select populations to keep certain genetic lineages alive. Cloning could help us maintain species in captivity until we can restore their habitats or add crucial genes back into a group of animals about to be released into the wild. Cloning won't be a cure-all, but given the state of the planet, it can't hurt to have options.

That's why frozen zoos represent the ultimate safety net, a genetic savings account for the future.* A century from now, scientists might have cloning down pat, or they might have an even better way to bring these cellsicles back to life. One tantalizing possibility involves stem cells, which can morph into any of the body's specialized cell types. Take a stem cell from an African wildcat and you may be able to coax it, in the laboratory, to grow into brand-new eggs or sperm. Scientists have managed to take the frozen skin cells of two highly endangered species—the white rhinoceros and the drill (a monkey)—and transform them into stem cells. The next step is to turn these cells into sperm and eggs, and then create test-tube rhinos and drills. The approach may turn out to be more efficient than cloning or a better option for species in which nuclear transfer has proved difficult. What's more, because using eggs and sperm to create new embryos leads to genetic remixing, it will yield rhinos with new combinations of genes and should be a better way to maximize genetic diversity. This stem cell work is still in early stages, but Dresser is thrilled by the promise it holds. "I may not live long enough to see some of the stem cell stuff applied to tigers or lions or elephants," she says, "but that's okay, because somebody had to start that process."

After spending years pioneering so many lab techniques, Dresser is passing the torch to the next generation of researchers, moving out from behind the microscope and into a more public role. She wants to make a forceful case for the need to develop reproductive technologies for endangered species—and to do so before it's too late. She's been traveling all over the country and batting ideas around with other experts. She's visited labs that specialize in livestock breeding and talked with scientists about how their research might be applied to more exotic animals.

* Ehrenfeld did endorse frozen zoos, writing that DNA banking "entails low risk, and seems worthwhile insurance against future discoveries and needs that we cannot know at this time."

The technology is moving quickly, and wildlife biologists have to stay on their toes. Breakthroughs in any number of fields—livestock breeding, companion animal medicine, human reproductive technology—can spark strategies for saving endangered animals. Even advances in computing and electronics could play a role: Just ask the brigade of biologists who are fighting for threatened species with an arsenal of high-tech tracking devices.

5. Sentient Sensors

In the idyllic decades after the end of World War II, Americans discovered nature. In the 1950s and '60s, families eager to see a slice of wildness flocked to places such as Yellowstone National Park. By then, the park's gray wolves had already disappeared, but there were other fearsome predators for visitors to see. Grizzly bears were a particular draw—so much so that the park's management installed bleachers at several garbage dumps so tourists could sit and watch the bears paw through food scraps and other trash. As park attendance rose, so did the number of human-bear interactions; the bears crashed through campgrounds looking for food and even outright begged for it at the side of the road. Unfortunately, these encounters didn't always end happily—for members of either species. The bears destroyed property and injured visitors, and rangers killed animals with a history of problem behavior.

Enter John and Frank Craighead, biologists and twin brothers who thought that learning more about the bears' lives could help Yellowstone's management reduce these interspecies conflicts. So the men decided to harness recent advances in radio and transistor technology to conduct some grizzly surveillance. Beginning in 1961, the Craigheads trapped bears at Yellowstone, tranquilized them, and then outfitted them with collars containing radio transmitters. (In case you're wondering: To trap a bear, try bacon, pineapple juice, or, of course, honey.) The stunned bears then continued on their way, but by using a radio receiver to tune into the signals issuing forth from the grizzlies' thick necks, the Craigheads were able to follow the bears as they ambled through the wilderness. As Frank Craighead recounted in his book:

> Beep, beep, beep, full of portent and meaning, the repetitive metallic pulse came in loud and clear on the crisp fall air. The sound had nothing of wildness about it. No deep primitive instinct of the chase stirred in us at the sound, nor did it evoke a feeling of oneness with nature. Yet this beeping coming to us in the vastness of Hayden Valley thrilled us as few sounds ever had. The vibrant pulsing signal, though new to the Yellowstone wilderness, told us that we were in communication with the grizzly we identified as bear Number 40, just as surely as the distant honking told us that the Canada geese were on the wing. But the beep was more specific than the honk of the goose or the guttural caw of the raven, for it emanated from one particular grizzly bear somewhere within the three thousand square miles of the park.

The technology opened up a whole new way of interacting with the wild world, and the Craigheads' project—one of the first large-scale uses of radio collars—signaled the birth of the modern era of wildlife tracking.*

* Among other things, the tracking data led the Craigheads to conclude that the park would be well advised to gradually phase out its open-pit garbage dumps, which attracted hungry bears.

Radio transmitters weren't much use to the era's marine biologists, in part because radio waves don't travel well through salt water. But these scientists didn't want to be left out of the tracking revolution that the Craigheads and others were launching on land, and during the 1960s and '70s, they started developing their own instruments. The first attempts were slapdash; one scientist measured the diving behavior of a Weddell seal with a pressure gauge and a wind-up kitchen timer. But biologists and engineers stuck with it, eventually creating devices that recorded information about marine mammals' dives over the course of days and months. They also started following fish using acoustic tags, which emitted sound waves that could be detected by underwater microphones mounted to boats. The sound waves, alas, didn't travel very far, so scientists had to trail fish closely in order to stay in range.

Over the following decades, advances in computing made wildlife tags smaller and more powerful. The development of satellite technology presented exciting new options; tags that communicate with satellites allow biologists to sit comfortably in their labs while zeroing in on a distant animal's exact location on the globe. We now have a burgeoning supply of sophisticated electronic tags, some smaller than a jelly bean, that can keep tabs on wild animals for months or years at a time. These devices are proving to be especially valuable for learning about life in the ocean; marine biologists can't go sit in the middle of the sea and watch the fish stream by the same way that Jane Goodall peered through the thick forests of Tanzania to study her beloved chimps. By bolting a tracking device to a shark's fin or implanting one in a tuna's belly, we landlubbers are gaining intimate access to the lives of ocean animals.

And not a moment too soon, considering that our oceans are in crisis. Heavy fishing, pollution, and climate change are all making life difficult for the species that dwell in the sea. Populations of marine animals—fish, mammals, reptiles, and birds—have declined by an average of 89 percent from their historical highs. The latest generation of electronic tags are a powerful weapon in the battle to keep wildlife healthy and thriving, particularly for the marine biologists whose subjects are so slippery.

Between 2000 and 2009, for instance, a team of California scientists used a slew of electronic tags to follow the movements of 1,791 marine animals from 23 different species. The venture, known as the Tagging of Pacific Predators (TOPP) project, helped researchers discover new migration pathways and marine hot spots—Goldilocks-like "just right" regions of the ocean where many species converge.* "When we start to understand how animals use the environment," says Randy Kochevar, a marine biologist at Stanford University who was one of the principal investigators of the TOPP program, "it puts us in a much better position to make informed decisions about how to manage and protect those populations."

TOPP was a hugely ambitious demonstration of the potential of marine tagging, but it was also a mere jumping-off point. TOPP has morphed from a local endeavor into an international one (called Global Tagging of Pelagic Predators, or GTOPP), and scientists are constantly dreaming up new tracking projects. As the latest generation of tagged animals go about their daily lives, the computers fastened to their bodies are doing more than simply recording their movements—they're also collecting data about the ocean and its changing conditions. In this way, electronic tags are shifting animals' roles from passive research subjects to active creature collaborators—and, perhaps, partners in saving their own watery worlds.

* TOPP was one of seventeen projects launched in 2000 as part of the Census of Marine Life, a massive ten-year global collaboration among 2,700 scientists in more than eighty different countries. The goal was to document the variety of life-forms that live in the world's oceans, from plankton to mako sharks, in habitats ranging from coral reefs to deepwater vents.

For us two-legged, land-walking, air-breathing brutes, it's all too easy to overlook ocean life. I know I have. In all my years chowing down on spicy tuna rolls, I had never—not once—stopped to consider the animal on my plate. But standing in the Tuna Research and Conservation Center (TRCC) in Monterey, California, it's all I can think about. The center, jointly operated by Stanford University and the Monterey Bay Aquarium, is essentially a big warehouse, and most of the floor space is taken up by three large round tanks. Resembling enormous kiddie pools, they are filled with 150,000 gallons of seawater and dozens of bluefin tuna.

It's no wonder I've got Japanese food on my mind. Bluefin have a bright pink flesh that is highly coveted for sushi and sashimi, and the fish can fetch staggering sums. (In 2012, a 593-pound specimen sold for $736,000 at a Tokyo fish market—more than $1,200 a pound.) This is the first time I've seen living bluefin, and they are magnificent animals, beefy and muscular, and yet, somehow, lithe. Silver and glistening, they look like enormous bullets. They thrash their tails back and forth with such energy that their tanks quake, and choppy waves travel across the water's surface.

These big bruisers are just babies, two and three years old; bluefin can live for thirty years and grow to be thirteen feet long and 2,000 pounds. They are strong and fast, able to reach speeds of 45 miles per hour and traverse entire oceans in a matter of weeks. (Tuna have huge geographic ranges, spending time everywhere from South America to Norway). Their fins retract, giving them exceedingly streamlined bodies, and they are warm-blooded, which makes them oddities in the fish world but keeps them toasty as they cruise through icy waters.

Bluefin tuna swim so fast, far, and deep that it has been difficult to learn about their lives in the wild. Marine biologists use satellite transmitters to track sharks, seals, and turtles, which spend time near the ocean surface, but tuna live beyond the reach of satellites.* So scientists had to develop an alternative solution. In the 1990s, they realized that they could take advantage of the fact that tuna are commercially harvested and outfit the fish with tags that store location information for later, rather than transmitting it in real time. The idea was that when a fisherman landed a tuna equipped with one of these "archival tags," he could remove the device and return it to researchers. The fisherman would get a financial reward for his service, and the biologists would get weeks, months, or years of detailed data that would enable them to reconstruct the tuna's path.

Barbara Block, a Stanford marine biologist who directs the TRCC, helped pioneer the archival tagging of tuna, and she has used hundreds of the devices to follow the migrations of fish in the Atlantic and Pacific. To deploy the tags, Block and her team head out to sea, where they often brave stormy weather as they fish for tuna that can outweigh them by hundreds of pounds. Once they've wrestled one of these giants into the boat, they lay it on the deck, cover its eyes with a wet towel, and use a hose to irrigate its gills with seawater. One team member makes a three-to-four-centimeter incision in the tuna's side and places an archival tag inside the abdominal cavity. The tag is a marvel of miniature engineering. It crams a multitude of electronics into a small stainless-steel cylinder approximately the size of a tube of lipstick. It contains a suite of en-

* The TOPP team, for instance, affixed satellite tags—each about the size of a deck of cards and equipped with a short antenna—to the dorsal fins of mako and blue sharks. Thereafter, whenever the toothy predator's fin slices through the surface of the water, the antenna is exposed and begins transmitting information to a network of satellites. The satellites triangulate the signal, determine the shark's approximate location, and send this information off to scientists. When the shark slips back below the surface, the apparatus switches off.

vironmental sensors, a microprocessor, a tiny battery, and enough memory to store years' worth of data.* The whole thing weighs in at one-tenth of a pound and can operate more than a mile below the ocean's surface, at temperatures below freezing. Tucked inside the tuna's belly, the tag will measure the fish's depth and internal body temperature as it swims.

When the researchers sew the fish back up, they leave the tag's "stalk," a long, thin tube attached to the metal cylinder, jutting outside the tuna's body. This stalk contains sensors that will measure the water temperature and level of ambient light as the fish steams across the ocean. The scientists also attach a brightly colored "streamer tag" to the outside of the fish, which will alert fishermen that there's a bounty on the electronic device hidden inside. The lucky fishermen who end up with these tuna in their boats can remove the implants, contact the scientists, and return the tags for payouts of as much as $1,000 per fish. ("Big $$$ reward," as the streamer tag says.) All this poking, prodding, and tagging takes less than three minutes. The team then pushes the fish out of the boat's "tuna door," sending it gliding down a wet blue tarp, Slip 'n Slide–style, into the ocean.

It could be weeks, months, or years before the fish is caught and the tag makes its way to Block and her colleagues. Once they have the device in hand, the scientists download all the data it's been collecting. They use a combination of readings, including those from light, water temperature, and time sensors, to calculate the fish's latitude and longitude on a given day. By stringing these locations together over the course of many days, they make highly detailed

* The transmission of an animal's movements in real time—through the radio collars on bears or the satellite transmitters on sharks—is generally referred to as biotelemetry. The use of instruments that store data, rather than transmit it instantaneously, is known as biologging.

maps of each tuna's aquatic wanderings. Block constructs these tuna trails for a variety of projects and programs. Since the mid-1990s, she has been tracking Atlantic tuna under the auspices of a research-and-conservation program known as Tag-A-Giant. For a decade, she tracked Pacific tuna for TOPP, and she now heads its successor, GTOPP. But all roads lead back to the TRCC, where Block and her colleagues study tuna biology, test out new tagging technology, and refine the techniques they use in the wild.

As I tour the facility, I run into Alex Norton, the facility's scruffy blond, visor-clad tuna manager. He holds his elbow out to me expectantly. Since the staff here often have wet, fishy hands, he says, the mode of greeting consists of elbow bumps. I angle my funny bone toward Norton and officially make his acquaintance.

Norton tells me that I've arrived just in time for a tuna feeding, and he enlists me to help. I don some gloves and climb up a ladder to a plank suspended just below the ceiling. I crouch and shimmy down the plank until I'm squatting directly over a tank of tuna. Norton follows behind. We start doling out the multicourse meal, dropping the offerings into the pool below us. First up, an *amuse-bouche* of vitamins, then a main course of squid and, for dessert, a tuna favorite: a bucket of fatty, oily sardines. It is a true feeding frenzy, the tuna zooming to the surface to snatch up the proffered delectables.

I ask Norton how a self-described "surfer dude" who waxes poetic about the beauty of the musculature of a tuna in motion feels about implanting electronics inside such impressive marine specimens. He says he doesn't think the instruments themselves physically harm the fish, but he has imagined what the psychological experience of being tagged must be like for a tuna. "You think of it as this alien-abduction scenario," he says. "You're swimming along and you eat something that looks wonderful. All of a sudden you're dragged toward this big giant thing—you know, it would be like a tractor beam, pulling you in—and then you go onto the mother ship, where they probe you, insert something, and chuck you back!"

It doesn't sound like a pleasurable experience, and tagging and tracking have long attracted controversy. In the 1960s, for instance, wilderness activists raised philosophical objections to the Craigheads' bear-tracking project. To these critics, radio collars represented an unwelcome human intrusion into the natural world. Other activists of the era were more concerned about animal welfare, worrying that big, bulky transmitters would cause discomfort, irritation, and pain.

Although tracking devices have evolved considerably since the 1960s, scientists still grapple with the effects of their instruments. Making even a small change to the body of a wild creature can have a big impact on survival and reproduction. In some studies, for example, penguins wearing time-depth recorders or radio transmitters took longer to find food and had higher rates of chick mortality. Researchers speculate that the devices interfered with the birds' streamlined silhouettes, increasing drag while they were swimming, and thus the amount of energy they had to expend. In certain species of fish, tagging has been associated with slower swimming speeds and growth rates, as well as muscle damage and scale loss at the site of attachment.

Surgically implanted tags can cause pain or lead to infections, while external ones can cause sores; biologists have documented cases in which the harnesses used to attach transmitters to sea turtles caused abrasions and tissue damage. Tracking devices can also attract predators, alter an animal's social status, or make it less desirable to potential mates. Poorly placed tags can snag on trees or brush and interfere with an animal's ability to swim, walk, or fly. Simply being caught and handled by humans can be traumatic, causing spikes in heart rate, respiration, body temperature, and the production of stress hormones, and leave animals susceptible to various diseases and pathogens.

These possibilities are problematic for animal welfare reasons, but also for scientific ones. We're tracking animals to learn more about them, and if the tag itself alters behavior, physiology, or survival, the data will be distorted, if not totally useless. So biologists who want to minimize the effects of tracking devices have to carefully consider countless variables. They must think through the physical and behavioral characteristics of an animal when deciding what kind of tag to use, where to place it, and how to attach it. Where, when, and how an animal is caught, restrained, handled, and released also matter. Some tags may be totally innocuous—for every study documenting the devices' adverse effects, there's another that shows tagged animals doing just dandy—but an ill-conceived one could be a death sentence.

It's not easy to perform controlled, long-term studies on how tags affect animal welfare, since it's difficult to get data on untagged animals as a point of comparison. So Block and her colleagues have tested out different tag shapes, attachment strategies, and surgical techniques with captive tuna. They implanted archival tags in tuna living at the TRCC and monitored the fish for months. The wounds healed well, and the only noticeable side effect was some "minor irritation" where the light-sensing stalk protruded from their bodies.

For all the discussion of how tags can harm animals, there's not much talk about how such devices could benefit them. Exhibit A: TurtleWatch, a program designed to protect loggerhead turtles, the giant, long-lived reptiles classified as "endangered" or "threatened" in the Pacific, Atlantic, and Indian oceans. The loggerheads that live in the northern Pacific nest in Japan and Australia but make yearly migrations across the open ocean, using their brown-and-white speckled flippers to paddle their way to the shores of the Golden State. Though the turtles aren't commercially harvested, they can swallow hooks or get tangled in lines set out by fishermen.

This "bycatch" of loggerheads is a problem for both the turtles and the fishermen. Federal regulations stipulate that the longline swordfish and tuna fishery operating around Hawaii cannot accidentally hook more than seventeen loggerheads annually. That's a collective total for all the boats working in the area—after someone snags the year's seventeenth turtle, all the fishermen must return to shore for the rest of the calendar year. In 2006, the fishery reached this limit unusually early—in March—and had to cease operations until the next year.

After that season, which was catastrophic for the fishing industry, Jeffrey Polovina and Evan Howell, both oceanographers at the National Oceanic and Atmospheric Administration Pacific Islands Fisheries Science Center, established TurtleWatch to reduce the turtle bycatch. Polovina, Howell, and their colleagues had already used satellite transmitters to track young loggerheads, and they'd discovered that the reptiles preferred water in a narrow temperature range: between 63.5 and 65.5 degrees Fahrenheit. The loggerheads also spent most of their time cruising around a region of the Pacific where large systems of marine currents converge; all sorts of buoyant, gelatinous critters pile up amid this churn and swirl of seawater, providing easy pickings for hungry turtles.

Polovina and Howell decided to use this information to predict where turtles might be on a particular day and encourage fishermen to avoid that area entirely. Since December 2006, that's what they've been doing under TurtleWatch. Every day, Howell examines the latest data on sea surface temperature and ocean currents and produces a map of the fishing grounds, using thick black lines to mark off regions where conditions will be especially turtle-friendly. The maps, which are produced in English, Vietnamese, and Korean, advise fishermen to avoid setting their lines in these areas and are dispatched daily to fisheries managers and individual boats. Since the program began, the fishery has never hit its maximum number of loggerhead encounters.*

The TurtleWatch approach doesn't make sense for species that fishermen *want* to catch, such as tuna. But there are ways that we can use tracking data to help protect the overexploited tuna populations. Since the early 1980s, fishermen have been subject to strict quotas; the International Commission for the Conservation of Atlantic Tunas (ICCAT) sets limits on how many pounds of bluefin can be pulled out of the water each year. ICCAT manages the Atlantic bluefin population by literally drawing a line down the middle of the ocean and treating the fish on each side as a distinct population. To the west of the line are tuna that breed in the Gulf of Mexico, while the eastern population breeds in the Mediterranean Sea. The western population, which has declined by more than 90 percent since 1970, is much smaller than the eastern one, so the quotas on the American side of the Atlantic are much more stringent.

It's a reasonable system, provided the fish stick to their side of the ICCAT line. "Well, when we started tagging and tracking bluefin tuna," says Randy Kochevar, the Stanford marine biologist, "one of the first things we realized is that nobody told them about this line down the middle of the ocean." Kochevar works in Block's lab, where researchers have been following the trails of Atlantic bluefin for more than a decade. Their tracking data reveals that in the spring and summer, the fish do indeed segregate themselves—a tuna born in the Gulf of Mexico will return to the Gulf to breed. During the rest of the year, however, the fish use communal foraging grounds

* A review of the 2007 fishing season suggests that the map predicted turtle locations reasonably well. Eight of the twelve loggerhead interactions that year happened when fishermen ignored the map and set their lines in the high-risk zone.

spread across the Atlantic. And as soon as the western tuna cross over the invisible ICCAT line, they can be harvested at a much higher rate. This finding helps explain why western tuna populations aren't bouncing back and points the way to better management plans. Block's team, for instance, has suggested establishing a new ICCAT zone, in the shared central Atlantic foraging grounds, governed by a strict catch quota. In this way, data from tuna tracking studies could be used to craft fisheries plans that lead to real recovery.*

As marine tracking matured, oceanographers realized that they could piggyback on biologists' tagging projects to learn about the sea itself. That's what happened when Michael Fedak, a marine biologist at the University of St. Andrews in Scotland, started tagging southern elephant seals. The blubbery behemoths—males can weigh in at more than 4,000 pounds—spend their lives in one of the most inaccessible places on the planet, enjoying winter in the frigid Antarctic waters. Some of the deepest divers on Earth, the mammals can descend more than a mile beneath the surface to hunt for dinner. The seals spend a few months every year on the beach, where they molt and breed, but when they slip back into the water, Fedak says, "they might as well be going off to another galaxy."

Eager to learn more about these seals' habitats, Fedak outfitted the animals with tags that would measure the basic physical characteristics of the water in which they were diving. Between 2003 and 2007, Fedak and his British, French, Australian, and American collaborators glued multifunction tags to the hairy heads of 102 ele-

* Tuna aren't the only commercially harvested species that could benefit from long-term tracking studies. As of 2010, 28 percent of marine fish populations were overexploited, and another 53 percent were being harvested at their maximum sustainable rate. Tags might help us discover better ways to manage these species, too.

phant seals.* Whenever an elephant seal dove beneath the surface, the gadget whirred away, measuring the water's pressure, temperature, and salinity at regular intervals. When the seal surfaced, the tag's satellite transmitter sent the data back to the lab. According to Fedak, the sensors in the tag are essentially identical to those that oceanographers lower into the sea from a ship, "except stuck on something hairy and warm."

In fact, as the numbers started trickling in, Fedak realized that oceanographers were eager to see the information his seals were collecting. "These guys needed this data for this much grander job of understanding how the ocean behaves," he says. Oceanographers are now using the temperature, salinity, and pressure readings from the seals' deep dives to construct detailed profiles of entire vertical columns of water. Because the animals routinely plunge under ice caps, where ships can't navigate, they are illuminating parts of the planet that have, until now, been complete blind spots. Among other things, tagged elephant seals have revealed previously undiscovered troughs at the bottom of the Antarctic Ocean. These valleys, which can funnel warm water under ice caps, may explain why some ice shelves have been melting faster than expected. Today, marine mammals have collected 70 percent of the Antarctic Ocean data in the World Ocean Database, and the U.S. Integrated Ocean Observing System is working to incorporate data collected by all sorts of tagged swimmers into its models of ocean conditions.

Ice melt is just the beginning—global warming is raising water temperatures and levels, and changing its acidity and salinity. Experts are also predicting long-term changes in precipitation, storm frequency, and ocean currents and circulation. These shifts are already having profound effects on marine life. As waters warm,

* The glue and tags stay put until the seals undergo their annual molt and then simply fall off, leaving no lasting scars.

many species of fish are moving toward the planet's poles, and there have been shifts in the distribution and availability of various nutrients and food sources, including plankton, the floating organisms that are central to many marine food webs. Scientists have, in turn, linked changes in prey availability to findings that porpoises are taking longer to mature, seals are giving birth later in the year, and whales are having fewer calves. Of course, some species are adapting to our warming world, but those that fail to adjust quickly enough could find themselves staring down the barrel of extinction.

Data from tagged elephant seals and other marine animals will help us monitor, forecast, and prepare for the drastic environmental shifts that threaten ocean life and predict what will happen to animals as the seas change. For example, scientists have used tags to estimate a seal's buoyancy, an indirect measure of body fat. A fat elephant seal is a thriving, well-fed elephant seal, and by using buoyancy, location, and other tag data, scientists can construct maps of where elephant seals find food and what ocean conditions are like there. "We can then run models around where those kinds of places might be in the future and how far away they are from where animals might breed," Fedak says. "It's the beginning of asking questions about how oceanographic changes might affect populations, of saying, 'Well, if things do shift . . . what will happen to the beasts?' " The latest generation of tags and sensors are turning elephant seals and other marine animals into more than scientific subjects. "We're making colleagues of the animals," Fedak says. "There really is an opportunity for us to understand the ocean not only for our reasons but for them as well. The animals and us are all in this together."

Tagging technology is advancing at a rapid clip, and tracking projects are proliferating. Several years ago, scientists launched the

Ocean Tracking Network, a $168 million project based at Canada's Dalhousie University. It's a collaboration of more than two hundred scientists in fifteen nations that aims to follow the movements of thousands of marine animals, from seals to eels, all over the globe. The project relies on acoustic tags, which emit pulses of sound that can be detected by underwater receivers. The basic technology has been in use for decades, but the Ocean Tracking Network is taking it to the next level by installing arrays of underwater "listening stations," capable of picking up the signals of any tagged animal that happens to swim past, along the ocean floor. The receivers, which are the approximate size of fire extinguishers, record the animal's presence, upload any data that's been stored in its tag, and relay the information to researchers. So far, OTN technicians have set up hundreds of these receivers on the seabed off the Canadian coast, with smaller deployments near Australia and South Africa. The goal is to establish similar arrays in all the world's oceans.

New kinds of tags are providing even more detailed information about the daily lives of ocean animals. A team of Hawaiian biologists, for example, gave Galápagos sharks electronic "business cards," acoustic tags capable of detecting other tagged sharks and recording when the predators encountered one another in the wild. Widespread use of these devices could help us learn more about how different individuals and species share the marine environment. A number of other labs are using tags that measure acceleration to determine when a shark is mating or a sea lion is hunting for fish.

Scientists who track deep-sea fish are beginning to deploy a second kind of device, known as a "pop-up" satellite tag, alongside their archival instruments. When attached to the outside of a fish, these pop-up tags collect and store the usual information about temperature, light, and depth. After a predetermined number of days, the tag automatically detaches from the fish and floats to the surface, where it sends its stored data to satellites. These tags are bigger, heavier, and more expensive than archival implants, and because of slow transmission speeds, they can transmit only small amounts of data. But prices and sizes are dropping, and the technology is being used on a variety of large fish, including swordfish, marlin, and tuna. (Barbara Block, who piloted the use of these devices with bluefin tuna, employs both pop-up tags and archival ones in her tracking studies.)

As electronic tags shrink to near invisibility, it's becoming possible to track an ever-expanding menagerie of marine and terrestrial species. A Canadian company sells a radio transmitter that is smaller than a fingernail and is practically weightless, at 0.25 grams. In 2010, researchers reported that they had used miniature tags to follow iridescent orchid bees as they flew through the tropical forests of Panama, and a group of Swedish scientists have shown that we may be able to track the movements of water fleas (*Daphnia magna*)—millimeter-sized, freshwater crustaceans—by attaching fluorescent nanoparticles to their tiny little shells.

Etienne Benson, the author of *Wired Wilderness: Technologies of Tracking and the Making of Modern Wildlife*, expresses mixed feelings about these advances. "We're tracking everything," he says. "Almost everywhere you go there is a committee of scientists or wildlife managers that is trying to manage the world. I think we can ask questions about what kind of world it is we're creating where we want to manage and keep track of everything. All the time." Benson, a research scholar at Berlin's Max Planck Institute for the History of Science, says electronic tags appeal to us because they provide another way to bring the wild world under our control. The rise of the tracking devices in the first place was driven by the fact that, as Benson puts it, "Wildlife managers needed to make *manageable* wildlife." (The Craigheads' research, after all, was spurred by a desire to keep grizzlies and humans away from each other.)

While Benson acknowledges that tracking devices can generate valuable data, he wonders whether we're being seduced by our new tools: "Do we really think that if we put a tag on everything, we're going to resolve problems of living harmoniously in nature or having a sustainable world of resources?" he says. "It's a kind of utopianism: 'If we just get everything tagged, if we just get the right sensor network out there, then everything's going to work fine.'"

Clearly, just knowing the whereabouts of the world's animals is not, in and of itself, a solution. We still have to use the information in the right way, and political and economic considerations often derail conservation. But if we want to protect animals, the more information we have about them and their habitats, the better. Plus, even Benson acknowledges that tracking devices have real benefits in another regard: public engagement. Tags that communicate with satellites allow scientists to broadcast the whereabouts of free-ranging animals online, in real time, for all of us to see, giving wild creatures their own pack of paparazzi. By providing closer encounters, even virtual ones, with other species, our electronic tools are bridging the divide between humans and animals.

TOPP researchers, for instance, conducted their own elephant seal tracking project and posted the seals' whereabouts on a public website. There, an interactive map displayed each seal's individual journey in the Pacific. I started checking in on the animals, cheering on the males as they traveled up the Pacific coast and worrying as the seal moms slid off into the dark night, abandoning their month-old pups for good. It was high drama on the high seas, and I devoured updates as if they were Facebook posts from my closest friends. (As luck would have it, the TOPP team set up Facebook accounts for some of its seals.)

I developed a particular soft spot for a loser male the researchers had named Jonathan Sealwart. The elephant seal, I learned, is at the bottom of the social totem pole and sleeps all alone on a California beach. He has no harem of admiring lady seals and may go his entire life without mating. He certainly can't rely on his looks; he has what must be one of the world's most hideous animal faces, with a droopy proboscis that appears to be melting off his face. To add insult to injury, Jon Sealwart has far fewer Facebook friends than the seal named after his late-night Comedy Central colleague, Stelephant Colbert.

TOPP isn't alone; projects tracking everything from albatrosses to sea turtles have made the animal world accessible to us online at all hours of the day. "You see conservation organizations and scientists all trying to forge connections using these technologies," Benson says, "to really give people insight into the everyday lives of nonhuman animals in a way that wasn't possible before. That can be tremendously valuable."

Even the act of assigning animals proper names—something that usually goes hand in hand with following the travels of specific individuals—can help us form emotional attachments to them. (Consider that pets are named but laboratory animals almost never are.) Thanks to proper names, I could do more than learn about the general characteristics of elephant seals; I could forge a bond with Mr. Sealwart, a specific seal with a unique history and personality. As Sune Borkfelt, a scholar at Denmark's Aarhus University, wrote in a 2011 paper, "[G]iving an animal a name does often draw it closer to us." Assigning names to individual animals can remind us that they are sentient subjects of their own lives, rather than mere objects, and it can highlight what we have in common with other species, rather than what sets us apart. Coming to know just a few wild individuals could prompt attachment and affection for an entire species and make us more invested in safeguarding their habitats and their futures. Tagging-and-tracking technology is helping us learn more about marine animals and the risks they face, simultaneously making us *want* to protect these creatures and giving us the knowledge we need to actually do so. Jonathan Sealwart may be a loser in the world of seals, but thanks to a little electronic device glued to his head, he's got a gang of human friends—more than five hundred of them, according to Facebook—and we, at least, are rooting for him.

6. Pin the Tail on the Dolphin

Winter's life began with a phenomenal stroke of bad luck.

In December 2005, when the Atlantic bottlenose dolphin was just a few months old, she was swimming with her mother in Mosquito Lagoon, along central Florida's Atlantic coast. Somehow, she got herself tangled in a crab trap. An eagle-eyed fisherman spotted her struggling and called in a wildlife rescue team. They found the calf gasping for air, her heart racing. The volunteers gently positioned the dolphin on a stretcher, carried her out of the water, and drove her across the state to the Clearwater Marine Aquarium.

She was in bad shape when she arrived—exhausted, dehydrated, and sporting numerous cuts and abrasions. She could barely swim, and trainers stood in the tank with her, holding her little body up in the water. No one knew whether she'd make it through the night. But she was a survivor, lasting through those initial hours and the following days, too.

Slowly, with bottle feeding and round-the-clock care, the team nursed the calf back to health. As Winter began to stabilize, though, other problems emerged. A line from the crab trap had been wrapped so tightly around her tail that it had cut off the circulation. The tissue was necrotic: The dolphin's skin started peeling off, and the tail itself began to decay. One day, Winter's caretakers found two of her vertebrae at the bottom of her pool. Winter was getting her strength back, but her tail was clearly a goner. And what kind of future could there possibly be for a dolphin without a tail?

Though she didn't know it, in one way, Winter was lucky—she was born in the twenty-first century, and there has never been a better time for an animal to lose a body part. Materials ranging from carbon-fiber composites to flexible, shape-shifting plastics are making it possible for us to design artificial appendages for patients that fly, trot, or swim; prosthetists have succeeded in creating a new beak for an eagle, a replacement shell for a turtle, and a false foot for a kangaroo. Surgical techniques are enabling vets to give cats and dogs bionic legs that are permanently implanted in their bodies, and advances in neuroscience hold out the promise of creating prostheses that can be directly controlled by the brain.

Whereas affixing sensors and tags to animal bodies could help save entire species, artificial tails and paws represent the other end of the spectrum, a way to provide a (sometimes literal) leg up to unlucky individuals. Prosthetic devices aren't appropriate for every animal—indeed, one of the challenges prosthetists face is determining what's in the best interest of bodies that look nothing like our own—but when we get it right, our custom-designed and individually engineered devices are helping us aid animals one life and limb at a time.

If there's any place to begin an investigation into the power of animal prosthetics—and the challenges involved in creating them—it's the Clearwater Marine Aquarium. Home to dolphins, stingrays, sea turtles, otters, and assorted other marine creatures, the facility is located on an island just off Florida's Gulf coast. The bright blue building sits right at the harbor; on a sunny spring morning, a dozen small boats bob among the docks. Inside, cheesy but cheerful island music plays on an endless loop. A few stairs lead from the main lobby to an open-air deck, where two dolphins kibbitz around in a large tank. It's easy to pick out Winter—instead of a long, full tail, she has a little curled stump that hangs off her torso like a comma.

Even with her abbreviated tail, Winter looks at home in the water, gliding and playing just like her fellow cetaceans. She's adapted to her unique body by adopting some unusual swimming techniques. Dolphins typically use their pectoral fins for balance, but Winter "cheats" and uses hers as little oars. And without the pair of flukes that normally adorn the end of a dolphin's tail, Winter lacks a dolphin's normal system of propulsion. So she's taught herself to swim like a fish, moving her body from side to side, rather than up and down, as dolphins normally do.

Unfortunately, this fish-like swimming posture puts pressure on Winter's spine, causing it to curve unnaturally. And in the months after her rescue, Winter's caretakers began to worry that her strange method of swimming would cause permanent injury. In September 2006, an aquarium official mentioned this concern in an interview with National Public Radio, which was airing a segment about Winter. A prosthetist named Kevin Carroll happened to be in his car, radio tuned to NPR, when the segment aired. As Carroll listened to Winter's saga, he thought: *I could put a tail on that dolphin.* A prosthetic tail, Carroll believed, might prompt Winter to start swimming like a dolphin again and stave off a lifetime of disability.

Carroll grew up near a hospital in a small Irish town, and seeing the ailing and injured children come and go inspired an interest in fixing the human body. He trained in prosthetics in Dublin, visited the United States, and never left. Today, Carroll is the vice president of Hanger Prosthetics and Orthotics based in Austin, Texas, and one of the world's leading prosthetists. He is constantly on the road, crisscrossing the country as he outfits injured patients with artificial limbs. He consulted on the case of Oscar Pistorius, the double-amputee sprinter from South Africa, has worked with world-class mountain climbers, and regularly watches his patients compete in the Paralympic Games.

Carroll's main focus is on helping humans, but every once in a while, someone will walk into his clinic with a three-legged dog or a beakless bird and ask for his help. As an animal lover, Carroll finds himself unable to resist donating his weekends to the cause. Over the years, he has worked with his Hanger colleagues to make prostheses for a veritable menagerie of animals: dogs, ducks, sea turtles, "whatever comes our way," he says. "I've sort of become the Dr. Doolittle of prosthetics."

The aquarium agreed to let Carroll take a crack at a prosthetic dolphin tail, and he began assembling his team. He knew who he wanted for a partner: Dan Strzempka, a prosthetist in Hanger's Sarasota office. Strzempka, who has worn a prosthetic leg since he was run over by a lawn mower at age four, is a Florida native with a passion for the ocean and the creatures that live there. But he wasn't sure what to make of Carroll's proposal to take on Winter as a patient. "At first, I thought he was joking," Strzempka recalls. "Or I thought he was crazy." As soon as he realized that Carroll was serious, Strzempka decided he was up for the challenge. One way or another, the men would give that little dolphin a tail.

Carroll and Strzempka have agreed to meet me at the aquarium and walk me through how they tackled the task. They are an odd pair; Carroll is slight and cue-ball bald, with a white beard, while Strzempka is tall, tanned, and solidly built. The staff beam when they see the prosthetists arrive, embracing them like family, and we slowly make our way to Winter's pool, stopping to greet more people every few feet. When we finally get to the dolphin tank, Strzempka leans up against the railing and calls to Winter: "Hey, girl! What's up, buddy?"

"Good marnin'!" Carroll shouts out to her in his Irish brogue.

Over the past five years, the men have spent countless hours standing here beside this tank. Winter is unlike any other patient they'd ever treated before, so their first task was understanding her body. Carroll and Strzempka began a crash course in dolphins, reading up on their anatomy and physiology and watching slow-motion videos of the cetaceans swimming to understand their biomechanics. Though animal prosthetists can draw upon human medicine, success often requires a degree of ingenuity; knowing how to build a leg for a human amputee won't get you far if you want to replace an elephant's missing foot or outfit a dog with a faux paw. So prosthetists often find themselves MacGyvering each animal appendage, custom designing and individually engineering it. Sometimes they end up inventing materials or techniques that have never been used in prostheses before.

In Winter's case, the basic plan seemed easy enough—Carroll and Strzempka decided to create a plastic tail that would slip over what remained of Winter's peduncle, the muscular back half of a dolphin's body that normally runs from the dorsal fin to the tail flukes. The challenge, they realized, would be figuring out how to keep the prosthesis on. Winter would be putting an incredible amount of force on the tail while swimming, but she wouldn't be pressing the entire weight of her body into it, as a human does with a prosthetic leg. "Water," Strzempka reminds me, "is a totally different environment." What's more, dolphin skin is slippery, sensitive, and delicate—and very easily injured.

Human amputees commonly use soft liners to cushion their

stumps and shield their skin, and Carroll and Strzempka decided that Winter would need something similar. But the standard human liner wouldn't do—for Winter, they'd have to create a brand-new material, soft enough to protect her skin, sticky enough to stay put on a slick surface, and strong enough to withstand daily use and abuse in a tank full of salt water.

They enlisted the help of a chemical engineer, who tinkered with the recipe for a gel liner common in human prosthetics, trying to create a version more suitable for a dolphin. The first few prototypes he made were promising, but their performance was inconsistent, and there were several dramatic failures, including a fire that burned a warehouse to the ground. ("It was a small warehouse," Strzempka assures me.) Finally, the engineer nailed it.

"It's incredible material," Carroll says, as we sit inside the trainers' office at the aquarium. He hands me a sheath of the rubbery gel, which is white, jiggly, and slightly gummy to the touch. It resembles nothing so much as a supersized piece of calamari. Technically, the material is a thermoplastic elastomer—a mixture of plastics that begins as a liquid and can be molded into a variety of shapes when heated—but everyone just calls it the "dolphin gel." Eager to show off its properties, Carroll takes a two-foot strip of the dolphin gel and hands the other end to Strzempka. He starts walking backwards. Two, five, ten feet—the material just keeps stretching. Finally, Carroll lets go. His end whips back across the room. Strzempka holds up the gel; it looks as good as new, neither distended nor deformed. The men beam, and I get the sense that this is a well-rehearsed stunt. The gel also provides serious cushioning, which Carroll demonstrates by wrapping his hand in the liner and beating it furiously with a heavy mallet, before breaking into a grin and pulling out his unharmed hand.

To make sure that Winter didn't reject the strange material, the dolphin's trainers introduced it slowly, giving her a piece of the gel to examine, then gently touching her body with it, and eventually wrapping it around her entire stump. They repeated the process with the prosthesis itself, starting by attaching a tiny, lightweight contraption to Winter's peduncle, working up to larger and heavier devices.

Winter's an old pro now, happily wearing a full-size, anatomically correct prosthetic tail. To put the device on, a trainer balances on a platform suspended in Winter's tank. With one swift command, Winter gets into position, pointing her head down toward the bottom of the pool and sticking her peduncle up out of the surface of the water. A trainer rolls a sleeve made of the dolphin gel onto Winter's stump. Then comes the prosthesis itself, which Carroll and Strzempka carefully constructed after taking a series of three-dimensional images and scans of Winter's body. The prosthesis has a flexible, rubberized plastic "socket" that slips on over the gel liner, hugging what remains of the dolphin's peduncle. The socket tapers into a thin carbon-fiber strip, which is bolted onto a pair of fake flukes. Suction keeps the entire apparatus on.

Though the device is modeled on a dolphin's natural tail, it's made of all sorts of unnatural materials, and Winter has to be supervised while she's wearing it. Winter's caretakers need to make sure that the tail doesn't suddenly start to slip off, for instance, or catch on something in the pool, and that the metal pieces don't accidentally injure one of her dolphin playmates. So Winter doesn't wear the tail all the time. Instead, it's reserved for her daily therapy sessions, when trainers lead the prosthesis-wearing dolphin through a series of drills designed to build up her muscles and reinforce proper swimming posture. (During these sessions, the trainers also use gentle pressure to stretch and straighten the muscles in Winter's stump.) The artificial tail helps keep Winter's spine in proper alignment, and with it on, Winter does, indeed, flick her tail up and down, rather than from side to side. "It's just beautiful to see her swim with it," Carroll says.* Winter's scoliosis has improved since she started wearing the device, and Carroll hopes the prosthesis, combined with regular therapy, will help the dolphin lead a long, healthy life.

Despite the progress she's made, Winter will spend the rest of that life in an aquarium; a dolphin without a tail, or with a human-fashioned one, is not a great candidate for survival in the wild. There's no telling how her prosthesis would hold up to years of constant use, whether it might fall off or fall apart, and Winter will need continuing access to trainers to reinforce proper swimming posture and doctors to monitor her spinal alignment. She'll need prosthetists on hand to repair damage to the tail, as well as to make other refinements. In fact, Carroll and Strzempka are still making several new tails a year for Winter, who has not yet reached her full adult size, tweaking the design as her body changes and her muscles develop. They also dream of making more dramatic improvements to the prosthesis. For instance, Strzempka says he would love to figure out how to incorporate a vacuum device that pumps air out of the tail whenever Winter moves it up and down. The result would be an even tighter seal and a constantly self-adjusting prosthesis.

Winter's tail has earned her full-fledged celebrity status. There are books, video games, and documentaries about her, and in 2011, Warner Bros. released *Dolphin Tale*, a 3D movie based on her story. (The prosthetist, or "mad scientist character," as Carroll calls him, is played by Morgan Freeman.) The aquarium's website and gift shop is chock-a-block with Winter gear: T-shirts, postcards, magnets, and toy dolphins that are also missing their tails.

But Winter has become much more than a powerful marketing

* You can see Winter swim with your own eyes, via the live webcam on the aquarium's website (www.seewinter.com).

tool—she has also become an ambassador for prostheses. Children with artificial arms and legs regularly visit the aquarium, and many are invited into the tank with Winter. The encounter can do wonders for a kid's psyche, Carroll tells me. "The psychological aspect of it is just incredible, for a child [who's] lost a limb," he says.

Winter has helped human amputees in more concrete ways, too; as word spread about the so-called dolphin gel, prosthetists began ordering it for their human patients. The material, which grips the skin better than the liners commonly used with people, has proven especially useful for amputee athletes, whose replacement limbs start to slide off when they sweat. Strzempka, an avid golfer, became a convert the first time he tried the gel in his own artificial leg. "The stickiness is a huge benefit, especially in Florida," he says. "If you're golfing thirty-six holes a day, your skin becomes like a dolphin's—slippery." It didn't take long for Hanger to start selling "Winters-Gel" liners to everyone from seasoned triathletes to eleven-year-old girls. "Animals give back to us all so much," Carroll says. "We learn so much from working with them."

Injured animals easily capture our hearts, and it's natural to want to heal their wounds. Bringing home a bird with a broken wing or an ailing, tick-covered stray is practically a childhood rite of passage. (My stray was a sickly, starving Doberman puppy, who I found wandering in the Virginia woods. His skin was so wrinkled and loose that we named him Raisin.) We naturally empathize with critters that are suffering; some neuroscience research has revealed that the brain regions that are active when we see fellow humans in distress also light up like pinball machines when we see an animal in pain.

Animals have all sorts of ways of communicating their distress—they may cease to eat or groom themselves, or may pace, whimper, cry, or obsessively lick or rub parts of their bodies. Sheep in pain curl their lips, horses sweat excessively, apes and monkeys roll their eyes. Mice make grimaces, and scientists have developed a "mouse grimace scale" so researchers can assess their rodents' discomfort. But animals can also be incredibly "stoic," and since they can't talk to us, it's not always obvious whether they need medical attention.

Not everyone thinks that prosthetists are doing the right thing when they intervene in the lives of injured animals. Though Carroll and his colleagues are driven entirely by the desire to *help* their patients, human-designed, factory-manufactured appendages still represent a radical refashioning of animal bodies. And Carroll often encounters naysayers, other prosthetists or members of the public who insist that his devices won't work or that they'll cause wild creatures undue distress.

Some of the critics' concerns mirror those faced by the scientists who use electronic tags to track the movements of wild animals: Will this device cause physical or psychological discomfort? How will the animal adjust to having a foreign object attached to its body? Vets and doctors must weigh the answers to these questions against the potential medical consequences of *not* giving an animal a prosthesis. The Clearwater Aquarium could have spared Winter the medical scans, tail fittings, and training sessions that her prosthesis required, but the tradeoff might have been a lifetime of deformity and pain.

Not every case is so straightforward. When I visited Carroll at one of his clinics, he showed me photos and X-rays of a California sea lion missing part of its left flipper. The mammal's caretaker had just called, wanting to know whether a prosthesis was an option. Carroll ultimately decided to pass on the project because he thought the sea lion was doing just fine and wasn't sure a false flipper would improve her quality of life. But there's no way of knowing, for sure, whether that was the right decision.

And then there's even trickier territory: the use of prostheses to alleviate an animal's *mental* anguish. Take the dog owner Gregg Miller, for example, who swears that his beloved bloodhound Buck was downright depressed after getting neutered. According to Miller's recollection, Buck came out of surgery, went to clean himself, noticed his missing 'nads, and then looked up mournfully at his owner. "Good God, it was horrible," Miller recalls. In those awful first days after the operation, a novel thought occurred to Miller: Maybe he could buy some fake balls and use them to make Buck look whole again. "Don't they make artificial testicles so it can reduce my trauma in neutering Buck and Buck's trauma at losing a body part?" he wondered.

When Miller discovered that no one made prosthetic dog testicles, he decided to create them himself. "People thought I was nuts," he says. "No pun intended." Working with veterinarians over the course of two years, Miller developed "Neuticles," and launched the CTI (Canine Testicular Implantation) Corporation to sell them. The implants are shaped like oversized lima beans and are designed to perfectly replicate the "texture and firmness" of the genuine articles. (I'll have to take Miller's word for it.)

The first dog received his counterfeit gonads in 1995.* The Neuticles were popped in while the pooch was on the operating table having his real testicles removed, adding just a few extra minutes to the surgical procedure. When the dog came to, it looked as though he'd never even been neutered. (As CTI's slogan asserts: "It's like nothing ever changed.") The prostheses come in an assortment of sizes and materials; prices range from $109 for a "petite" pair of the original Neuticles to $1,299 for a set of extra-extra-large, top of the line NeuticlesUltraPLUS. The company also sells models designed

* The breakthrough was too late for Buck, who eventually died of liver cancer. "Even though he never got Neuticles, he changed the world," Miller says.

for cats, horses, and bulls, and more than 250,000 pets in forty-nine countries have now received fake balls.

That's a lot of animals that have been spared the humiliation of emasculation. Perhaps. It's hard to know what the dogs think of the implants, or whether they'd even notice if they suddenly vanished. That's the challenge involved in outfitting animals with prostheses: Other species can't weigh in on whether and how they want their bodies to be remade. Though brain imaging lets us witness animal minds in action, as one neural circuit or another lights up, we'll never truly comprehend what life is like, on the level of subjective experience, for a member of another species.* (We have enough trouble imagining what life is like in another *person's* shoes.)

In his famous essay "What Is It Like to Be a Bat?" the philosopher Thomas Nagel expounded on this very problem. As he wrote:

> It will not help to try to imagine that one has webbing on one's arms . . . which enables one to fly around at dusk and dawn catching insects in one's mouth; that one has very poor vision, and perceives the surrounding world by a system of reflected high-frequency sound signals; and that one spends the day hanging upside down by one's feet in an attic. In so far as I can imagine this (which is not very far), it tells me only what it would be like for me to behave as a bat behaves. But that is not the question. I want to know what it is like for a bat to be a bat. Yet if I try to imagine this, I am restricted to the resources of my own mind, and those resources are inadequate to the task.

* Since MRI machines require a subject to lie motionless inside a narrow tube for an extended period of time, it's been tricky to get good scans of animals that are awake. Gregory Berns, a neuroscientist at Emory University, recently showed that we can get dogs to be good MRI subjects through simple training. Using nothing more than positive reinforcement, Berns and his colleagues taught Callie, a two-year-old mutt, and McKenzie, a three-year-old border collie, to climb into the MRI

Neutering may well be traumatic. Surgery is stressful and recovery can be painful. A dog's gonads produce sex hormones, and removing them can cause behavioral changes, especially a reduction in mounting, marking, and aggression. But just because neutering alters sexual *behavior* doesn't mean that it causes a crisis of sexual *identity.* As the Humane Society explains in an online guide to spaying and neutering, "Pets don't have any concept of sexual identity or ego." And behavioral changes don't necessarily equal distress. Miller says that he finds neutering to be "a creepy, creepy thing. You're modifying your dog, you're kind of playing God." But Neuticles don't *unmodify* the animal—they merely add a second alteration on top of the first.

Though I have yet to find any peer-reviewed research on whether Neuticles can prevent neutering-related trauma in dogs, a study of monkeys provides a hint. Scientists studying the effects of monkey castration used Neuticles as a control—after removing the testicles of half the animals, they inserted silicon imposters. That way, all the male monkeys would continue to look identical to the other members of their social group. However, the prosthetic balls didn't prevent the neutered primates from behaving more submissively than their intact counterparts. The finding suggests that it's the absence of hormones, not some sexual identity crisis that results from looking like a eunuch, that causes behavioral changes in neutered animals. And Neuticles don't restore a male dog's normal hormone levels, nor do they spare him the trauma of the surgery itself.

So are Neuticles really for dogs? Or are they for humans, a way for us to atone for the castrating we put our pooches through? Most of his customers, Miller says, are "freaking out over neutering their dog." The possibility of testicle implants soothes their psychic pain,

---

tube, position their snouts on a chin rest, and then lie there, motionless, until the scan was complete.

and some animal welfare groups have endorsed Neuticles because they may spur pet owners who are on the fence about castration to go ahead and get their pets fixed.* (In fact, when Miller heard that President Bill Clinton had expressed hesitation about neutering Buddy, the First Dog, he did something that required both literal and figurative *cojones*; he asked the leader of the free world to give some thought to a prosthetic package.)

I never considered Neuticles when I got my dog, Milo, neutered, perhaps because I'm a woman. According to a survey of nearly sixteen thousand Australian dog owners, men are twice as likely as women to believe that neutering fundamentally changes a dog's "maleness."† It occurs to me that Neuticles might be a lot like truck nuts, those fake plastic testicles men sometimes affix to the back of their vehicles—there to telegraph the virility and manliness of the owner. Indeed, one male customer reported that his only disappointment with Neuticles was that he and his wife hadn't gotten their dog a bigger size. According to one scholar of gender studies and human-animal interactions, "many men continue to view their male pets as personifications of their own egos and libidos."

Though Neuticles are a bit, well, *nutty*, they don't strike me as cruel. Especially when we compare them with some of the other things we do to dogs' bodies. Take tail docking, for instance, in which the last several inches of a puppy's tail are removed, usually without anesthesia, sometimes with extremely crude instruments such as scissors or razors. The American Kennel Club (AKC), which develops guidelines used to judge canines in competition, prefers boxers,

* Neutering may be tough on dogs, but veterinarians and animal welfare groups overwhelmingly support the practice as a way to drastically reduce the number of unwanted pets in the world.

† For women, however, Miller has created a special line of Neuticle jewelry; for a small fee, the discerning lady can now wear real, 100 percent genuine, fake testicles around her neck.

rottweilers, cocker spaniels, and dogs belonging to dozens of other breeds, to have docked tails.* In other words, an ideal specimen is one that's been surgically reshaped by humans.†

Speaking of surgery, some vets are now giving dog owners the opportunity to have their pooches nipped and tucked. Pet plastic surgery can have a medical rationale—nose jobs can make it easier for some breeds (such as pugs) to breathe, face lifts can eliminate folds of skin that trap bacteria, and doggy braces can alleviate sores caused by crooked teeth. But one veterinary surgeon in Brazil says he has no problem performing the procedures for purely cosmetic purposes. "Why shouldn't a dog be beautiful?" he told the BBC. "Beauty is desirable. We all like talking to someone who looks good and smells nice. It's the same with dogs." But does the search for beauty alone, especially in the eye of a human beholder, really justify putting these canines under the knife?

Neuticles provide another way for us to project our own aesthetic ideals onto our pets, but the false balls also represent something more complicated than that. While few would argue that docking a cocker spaniel's tail makes the animal happier, hundreds of thousands of customers think synthetic testicles are good for their dogs. (One pet owner's silly silicone sac is another's medical miracle.) Neuticles—and the strange mixture of motives that may

* Some of the AKC's breed guidelines simply describe characteristics of healthy animals, explaining that certain dogs should have silky and glossy coats, or a smooth gait, or a full set of well-aligned teeth. But other stipulations reflect seemingly arbitrary aesthetic preferences. Consider the Labrador retriever, America's prototypical family dog. Labs that want a shot in the show ring better have black or brown noses. According to the AKC, "a thoroughly pink nose . . . is a disqualification." A Lab's eyes should be hazel or brown. "Black or yellow eyes give a harsh expression and are undesirable," the AKC says. As for coat color? "A small white spot on the chest is permissible, but not desirable."

† The United Kingdom has banned tail docking in dogs, with a few limited exceptions.

prompt their purchase—illustrate how hard it can be to untangle our own interests from what best serves an animal.

Even Carroll, who says his work is motivated purely by a love of other creatures, faces critics. But he is undaunted by those who say that animal prostheses are a waste of time and money. "We try to work around positive people [who] think there's options out there, that there are solutions that can help an animal get back up and walk again," he says. "I think it's critical to take care of our animals and to help rehabilitate them when they get injured. Mostly the ones that we see, it's humans that have injured them, and I think it's important that humans put them back together."

Carroll and his colleagues have worked with animals for whom prostheses have been lifesaving. After Winter's story appeared in a local newspaper, Carroll got a call from Lee Fox, who runs Save Our Seabirds, an avian rescue and rehabilitation facility in Sarasota, Florida. Sandhill cranes commonly come into Fox's care; the birds are often hit by speeding cars and wayward golf balls, irreparably damaging their delicate legs. Because of how the birds are built—with long, thin legs and big, heavy bodies—a crane with a bum leg is usually a crane that will be euthanized.

Fox was, she says, "truly, physically sick about having to put down one sandhill after another," so she began outfitting her birds with prostheses she jury-rigged out of some PVC pipes and sink stoppers. When she met Carroll, he fashioned a more comfortable, permanent solution, taking plaster casts of the legless birds and making prostheses out of lightweight plastic. The recipients included a crane named Chrisie, and Fox vividly recalls the day Carroll first put his device on the bird: "Chrisie walked in hers like it was her own leg." The birds even use their artificial limbs to scratch themselves, something healthy cranes do with their natural legs. As

Carroll notes, "Animals are wild, so they're very adaptable to their situation. We help them and they get along with their little lives."

Cranes aren't the only creatures that could be saved by a well-engineered false limb. Horses are usually put down when they break or seriously injure a leg, and while some dogs career around just fine on three paws, not all adapt as easily. A corgi, for instance, with its long, tubby body, can't manage on three legs, says Noel Fitzpatrick, a veterinary orthopedic surgeon with his own practice in Surrey, England. Fitzpatrick thinks vets have been too willing to euthanize pets facing the loss of a limb. "Animals deserve a good quality of life," he says. "I'm not saying that you shouldn't put a dog to sleep if there's no other option. But there are circumstances, lots of them, where prosthetics really should be used."

There's especially no excuse today, he explains, given the advances in veterinary and materials sciences that are giving injured creatures more options than ever. Feline and canine anatomy present unique challenges for prosthetists—just as Winter's body did. Though dogs and cats *can* wear devices that strap on to the outside of their bodies, they don't always do well with them. Below-the-knee amputations in these species don't leave much muscle or flesh for an external apparatus to grab on to, and their rounded bones are difficult for an artificial limb to grip. Above the knee, the animals have too much muscle, and lots of loose skin, surrounding their bones; even when a socket prosthesis is tightened all the way, cats and dogs can sometimes pull their legs right out. What's more, pets often kick, chew, or claw off a strap-on device.

Fitzpatrick has been pioneering an alternative kind of prosthesis that gets around many of these difficulties.* His approach, known as

* External prostheses have limitations even for human patients. As a boy, Fitzpatrick witnessed that firsthand. His uncle had a wooden leg, and one day, while they

osseointegration, involves implanting one end of a prosthetic leg in an animal's stump and fastening it to whatever bone is left. The metal implant then passes through the skin and can be attached to a specially engineered foot or paw, creating what Fitzpatrick calls "a bionic dog."

Fitzpatrick knew that osseointegration was not without risks: A bone-anchored prosthesis breaks the skin barrier, jutting out through a patient's stump, providing an easy entrance for bacteria and leaving the patient susceptible to serious infections. Fortunately for the vet, Gordon Blunn, a biomedical engineer at University College London, thought he had a solution to this very problem. Blunn believed that surgeons could learn from deer antlers, one of very few cases in nature where skin and bone are tightly bonded. The secret to this strong marriage, one of Blunn's graduate students discovered, is that antlers are covered with tiny pores. Collagen fibers grow into and through these openings in the bone, creating a permanent connection between the antler and the surrounding skin. A pore-pocked prosthesis that mimicked deer antlers, Blunn proposed, might interface with the skin to create a protective seal against infections.

Using antlers as inspiration, Fitzpatrick and Blunn designed a medical device for legless animals: the intraosseous transcutaneous amputation prosthesis, or ITAP. By 2007, they were using it in pets. One of the first patients was an American bulldog named Coal, who had a tumor in his front left paw. The normal treatment would have been full amputation of the leg, but Coal had arthritis in his other limbs and was likely to struggle as a three-legged dog. So Fitzpatrick agreed to give Coal an ITAP. With the dog under anesthesia, Fitzpatrick inserted a rod made of a titanium alloy into the center of what remained of Coal's radius. As the rod emerged from the

---

were sailing, Fitzpatrick accidentally knocked it overboard. "I watched it float away," he recalls. "I thought, 'This is crap.'"

bone, it opened up into what resembled an upside-down umbrella. Fitzpatrick stretched Coal's skin over this rounded cap, which was dotted with small holes, much like a deer's antler. Then the surgeon sewed the bulldog up, hoping that Coal's soft tissue would grow into and through the implant, creating permanent links between skin, metal, and bone. Fitzpatrick left a short segment of titanium jutting out from Coal's stump, into the open air. When the dog's wound healed, his owners could then pop an artificial paw on and off the metal rod.

The antler-inspired device worked—Coal's tissues gradually grew into the implant, forming a sturdy seal, and there were no signs of infection. The bulldog adapted beautifully to his new leg. "The ITAP didn't just improve Coal's quality of life, it gave him life," his owner wrote in a testimonial. "Coal led a perfectly normal life after the operation, there was absolutely nothing he couldn't do. He would use his prosthetic leg to bash at the door to be let out, and used it with his good paw to hold food and toys, as well as give you his 'paw' when he wanted a treat. The most amazing thing was people never noticed he had a prosthetic leg until they actually looked closely."

Coal has been followed by other successes, including a black cat named Oscar, who received two ITAPs after losing both his back paws in an unfortunate run-in with a mechanical grain harvester. Fitzpatrick, who was the subject of a documentary series called *The Bionic Vet*, estimates that he's done about two dozen of these animal ITAPs. The procedure has been so successful in animals that human trials of ITAPs are under way in Britain. (One of the first recipients was a woman who lost her arm during the July 2005 bombing of the London Underground.)

Fitzpatrick's technique is just one approach to osseointegration, and a handful of other vets are trying out their own variations. Denis Marcellin-Little, a veterinary orthopedist at North Carolina State

University, has a number of canine patients that are either overweight or very active and thus not good candidates for conventional, strap-on prostheses. With osseointegrated devices, Marcellin-Little has helped some of these formerly hopeless cases get back on their feet again. Meanwhile, one equine researcher is designing implants that could be used to build bone-integrated prostheses for horses; the devices might save the lives of racehorses such as Barbaro, the Kentucky Derby champ who had to be euthanized after breaking a leg and suffering from an escalating series of complications. (Veterinarians tried to repair Barbaro's leg, but it didn't heal well, causing additional strain on his good legs. Eventually, these limbs, too, began to break down, leaving the horse literally without a leg to stand on.)

Animal prostheses, and the research and development required to create them, are not cheap, and Fitzpatrick says he's sometimes asked why he's spending so much time and energy designing a leg for someone's dog. The endeavor is about more than just giving a pet a new limb, he says. "It's about life and love. It's about the bond of love between an animal and a human that is a small microcosm about how things could be better on Earth."

Osseointegration is also a step toward building an even more futuristic kind of prosthesis—one that is not only a permanent part of the body but also works seamlessly with the nervous system. With such devices, patients will be able to move manufactured limbs more like natural ones, wiggling carbon-fiber fingers or toes just by thinking about it. There have already been major breakthroughs. Monkeys outfitted with brain implants have been trained to use just their thoughts to move external robotic arms—in one case, using an arm to feed itself marshmallows—and paralyzed humans have performed the same feat. (Instead of eating marshmallows, the human volunteer used the robo-arm to give his girlfriend a high five.) Scientists at the Rehabilitation Institute of Chicago have succeeded with a different approach, taking the nerves that would normally control the movement of a missing arm and moving them into the muscles of the chest in several human amputees. Patients learned to use the nerve signals generated by the use of these muscles to control virtual arms on a computer screen, as well as motorized prosthetic limbs.

The main goal in developing these robotic limbs has been to improve the lives of human amputees or quadriplegics. In particular, the Defense Advanced Research Projects Agency (DARPA), part of the U.S. Department of Defense, has invested a lot of time and money in this area of research, hoping to find better solutions for injured veterans. However, just because animals aren't the *intended* beneficiaries of this work doesn't mean that they won't benefit along the way. Prosthetic innovations flow back and forth across species lines, and it wouldn't be surprising to see scientists refining nerve-integrated prostheses in injured animals or veterinarians borrowing tools and techniques being used in human patients. A variety of diseases and conditions cause hind leg paralysis in elderly dogs, for instance; by turning these canines into cyborgs, with brain-controlled prostheses, we could give them back control of their nervous systems.

Or we could do something even more extreme, building bionic animals and then taking control of their nervous systems ourselves.

7. Robo Revolution

In the 1960s, the Central Intelligence Agency recruited an unusual field agent: a cat. In an hour-long procedure, a veterinary surgeon transformed the furry feline into an elite spy, implanting a microphone in her ear canal and a small radio transmitter at the base of her skull, and weaving a thin wire antenna into her long gray-and-white fur. This was Operation Acoustic Kitty, a top-secret plan to turn a cat into a living, walking surveillance machine. The leaders of the project hoped that by training the feline to go sit near foreign officials, they could eavesdrop on private conversations.

The problem was that cats are not especially trainable—they don't have the same deep-seated desire to please a human master that dogs do—and the agency's robo-cat didn't seem terribly interested in national security. For its first official test, CIA staffers drove Acoustic Kitty to the park and tasked it with capturing the conversation of two men sitting on a bench. Instead, the cat wandered into the street, where it was promptly squashed by a taxi. The program was abandoned; as a heavily redacted CIA memo from the time delicately phrased it, "Our final examination of trained cats . . . convinced us that the program would not lend itself in a practical sense to our highly specialized needs." (Those specialized needs, one assumes, include a decidedly unflattened feline.)

Operation Acoustic Kitty, misadventure though it was, was a visionary idea just fifty years before its time. Today, once again, the U.S. government is looking to animal-machine hybrids to safeguard the country and its citizens. In 2006, for example, DARPA zeroed in on insects, asking the nation's scientists to submit "innovative proposals to develop technology to create insect-cyborgs."

It was not your everyday government request, but it was an utterly serious one. For years, the U.S. military has been hoping to develop "micro air vehicles"—ultrasmall flying robots capable of performing surveillance in dangerous territory. Building these machines is not easy. The dynamics of flight change at very small sizes, and the vehicles need to be lightweight enough to fly, yet strong enough to carry cameras and other equipment. Most formidably, they need a source of power, and batteries light enough for microfliers just don't have enough juice to keep the crafts aloft for very long. Consider two of the tiny, completely synthetic drones that engineers have managed to create: The Nano Hummingbird, a flying robot modeled after the bird, with a 6.5-inch wingspan, maxes out at an eleven-minute flight, while the DelFly Micro, which measures less than four inches from wingtip to wingtip, can stay airborne for just three minutes.

DARPA officials knew there had to be something better out there. "Proof-of-existence of small-scale flying machines . . . is abundant in nature in the form of insects," Amit Lal, a DARPA program manager and Cornell engineer, wrote in a pamphlet the agency issued to the prospective researchers. So far, nature's creations far outshine our own. Insects are aerodynamic, engineered for flight, and naturally skilled at maneuvering around obstacles. And they can power themselves; a common fly can cruise the skies for hours at a time. So perhaps, DARPA officials realized, the military didn't need to start from scratch; if they began with live insects, they'd already be halfway to their dream flying machines. All they'd have to do was figure out how to hack into insects' bodies and control their movements. If scientists could manage to do that, the DARPA pamphlet said, "it might be possible to transform [insects] into predictable devices that can be used for . . . missions requiring unobtrusive entry into areas inaccessible or hostile to humans."

DARPA's call essentially launched a grand science fair, one designed to encourage innovation and tap into the competitive spirit of scientists around the country. The agency invited researchers to submit proposals outlining how *they'd* create steerable insect cyborgs and promised to fund the most promising projects. What the agency wanted was a remote-controlled bug that could be steered to within five meters of a target. Ultimately, the insects would also need to carry surveillance equipment, such as microphones, cameras, or gas sensors, and to transmit whatever data they collected back to military officials. The pamphlet outlined one specific application for the robo-bugs—outfitted with chemical sensors, they could be used to detect traces of explosives in remote buildings or caves—and it's easy to imagine other possible tasks for such cyborgs. Insect drones kitted out with video cameras could reveal whether a building is occupied and whether those inside are civilians or enemy combatants, while those with microphones could record sensitive conversations, becoming bugs that literally bugged you.

As far-fetched and improbable as DARPA's dream of steerable robo-bugs sounds, a host of recent scientific breakthroughs means it's likely to be far more successful than Acoustic Kitty was. The same advances that enabled the development of modern wildlife-tracking devices—the simultaneous decrease in size and increase in power of microprocessors, receivers, and batteries—are making it possible to create true animal cyborgs. By implanting these micromachines into animals' bodies and brains, we can seize control of their movements and behaviors. Genetics provides new options, too, with scientists engineering animals whose nervous systems are easy to manipulate. Together, these and other developments mean that we can make tiny flying cyborgs—and a whole lot more. Engineers, geneticists, and neuroscientists are controlling animal minds in different ways and for different reasons, and their tools and techniques are becoming cheaper and easier for even us nonexperts to use. Before long, we may all be able to hijack animal bodies. The only question is whether we'll want to.

DARPA's call for insect cyborgs piqued the interest of Michel Maharbiz, an electrical engineer at the University of California, Berkeley. He was excited by the challenge of creating flying machines that merged living bodies and brains with electronic bits and bytes. "What I wanted at the end of the day was a remote-controlled airplane," Maharbiz recalls. "What was the closest thing to a remote-controlled airplane that I could get with these beetles?"

Maharbiz was an expert at making small electronic devices but an amateur when it came to entomology. So he started reading up. He figured that most scientists taking on DARPA's challenge would work with flies or moths, longtime laboratory superstars, but Maharbiz came to believe that beetles were a better bet. Compared with flies and moths, beetles are sturdy animals, encased in hard shells, and many species are large enough to carry significant cargo. The downside: Scientists didn't know much about the specific nerve pathways and brain circuits involved in beetle flight.

That meant that the first challenge was to unravel the insects' biology. Maharbiz and his team began working with several different beetle species and eventually settled on *Mecynorrhina torquata*, or the flower beetle. It is a scary-looking bug—more than two inches long, with fearsome claws and a rhinoceros-like horn on the forehead. Through trial and error, the scientists homed in on a promising region of the beetle brain nestled at the base of the optic lobes. Previous research had shown that neural activity in this area helped keep the insect's wings oscillating, and Maharbiz's team discovered that when they stimulated this part of the brain in just the right way, they could start and stop beetle flight. When they sent a series of rapid electrical signals to the region, the beetle started flapping its wings and readied itself for takeoff. Sending a single long pulse to the same area prompted the insect to immediately still its wings. The effect was so dramatic that a beetle in mid-flight would simply fall out of the air.

After he discovered these tricks, Maharbiz was ready to try building the full flying machine. The flower beetle's transformation began with a quick trip to the freezer. In the icy air, the beetle's body temperature dropped, immobilizing and anesthetizing the insect. Then Maharbiz and his students removed the bug from the icebox and readied their instruments. They poked a needle through the beetle's exoskeleton, making small holes directly over the brain and the base of the optic lobes, and threaded a thin steel wire into each hole.

They made another set of holes over the basalar muscles, which modulate wing thrust and are located on either side of the beetle's body. The researchers pushed a wire into the right basalar muscle. Stimulating it would cause the beetle's right wing to start beating with more power, making the insect veer left. They put another wire into the left basalar muscle; they would use it to steer the beetle to the right. The loose ends of all these wires snaked out of their respective holes and plugged into a package of electronics mounted with beeswax on the beetle's back. This "backpack" included all the equipment Maharbiz needed to wirelessly send signals to the beetle's brain: a miniature radio receiver, a custom-built circuit board, and a battery.

Then it was time for a test flight. One of Maharbiz's students called up their custom-designed "Beetle Commander" software on a laptop. He issued the signal. The antennae jutting out of the beetle's backpack received the message and passed it along to the circuit board, which sent electricity surging down the wire and into the beetle's optic lobe. The insect's wings began to flap. The empty white room the researchers used as an airfield filled with a buzzing sound, and the bug took flight. The beetle flew on its own—it didn't need any further direction from human operators to stay airborne—but as it cruised across the room, the researchers overlaid their own commands. They pinged the basalar muscles, prompting the beetle to weave back and forth through the room, as if flying through an invisible maze. It wouldn't have looked out of place going up against a stunt pilot at an air show. Another jolt of electricity to the optic lobe, and the beetle dropped out of the air and skittered across the tile floor.

As soon as Maharbiz presented his work, the news stories came fast and furious, with pronouncements such as "The creation of a cyborg insect army has just taken a step closer to reality," "Spies may soon be bugging conversations using actual insects, thanks to research funded by the US military," and more. A columnist speculated about the possibility of a swarm of locust drones being used as vehicles for launching deadly germs. There was chatter about beetles that had been "zombified," and references to "the impending robots vs. humans war."

When Maharbiz reflects upon this media frenzy, he admits that the immense public interest in his work doesn't surprise him. The research, after all, is practically primed to light up the futuristic-fantasy centers of our brains. Insects, even without modifications, seem like weird, alien organisms to many of us. As Maharbiz explains, "Insects have inherently some sort of strange, science fiction quality that a bunny doesn't have." Add in miniature electronics, flying devices, animal-machine hybrids, and covert military operations, and you have a recipe for dystopian daydreaming.

But Maharbiz bristles at the most sinister suggestions, at the media coverage that suggests his beetles are the product of, as he puts it, "some evil government conspiracy." As for the possibility that the U.S. government is planning to use the bugs to build a killer insect army or to spy on its own citizens? "I think that's nonsense," he says. His beetles haven't been sent out into the field yet—they still need some refinement before they're ready for deployment—but if and when they are, Maharbiz says he expects his bugs to be used abroad, in routine military operations. (Of course, some people may find that "equally reprehensible," he acknowledges.) There are civilian applications, too. Imagine, Maharbiz tells me, an army of beetle-bots, steered to the scene of an earthquake. The bugs could be outfitted with temperature sensors, guided through rubble, and programmed to send messages back to search teams if they detect any objects that are close to human body temperature; rescuers would then know exactly where to search for survivors.

Whatever the application, future insect commanders will have options that go beyond beetles. Maharbiz is working on a remote-controlled fly, which he anticipates being especially difficult to build. "The fly is so small and the muscles are so packed and everything's so tiny," he says, that even just implanting the electronics will be challenging. A Chinese research team has managed to start and stop flight in honeybees, and Amit Lal, the engineer who led the DARPA program, has created steerable cyborg moths.

One of Lal's innovations has been figuring out how to take advantage of morphogenesis, the process by which many species of insects transform from wriggling larvae to spindly, multilegged adults. During pupation, a baby insect wraps itself in a protective cocoon or shell while its soft, immature body becomes a more structurally complex adult one. (Lal's species of choice is the tobacco hawk moth, which morphs from a bright green worm into a brown-and-white spotted moth.) To Lal, this phase of the insect life cycle presented a unique opportunity; he hoped that if he inserted electronic components into a hawk moth when it was a wee pupa, the bug's body would rebuild itself around the implant. In one set of experiments, Lal and his colleagues pushed thin wires through the hard shell that protects a hawk moth pupa and positioned them in the insect's neck muscles and brain. Outside the bug's body, the wires linked up with a small circuit board, which the researchers left resting loosely atop the pupal case. They repeated the procedure with twenty-nine more pupae and then tucked them all away inside an incubator and allowed them to develop normally.

About a week later, the insects shed their shells, emerging as fully grown moths. Their bodies had in fact fashioned themselves around the implants; tissue had grown around the wires, securing them in place. The wires ran out of the moths' heads and partway down their backs, winding their way into the attached circuit board. All researchers had to do to begin steering the moths was plug their control system into the circuit board, a task that took a matter of seconds.

These kinds of pupal surgeries have much to recommend them, the researchers say. They lead to more stable, permanent interfaces between electronic devices and living tissue. The approach may also be less traumatic for the animals; bugs heal easily during pupation, and since the adults are born with circuit boards hanging out of their backs, they're less likely to perceive them as foreign objects or extra weight. (After all, the bugs will never know a life in which they *aren't* attached to circuit boards.) It's also much easier to operate on a pupa than an adult insect. The procedure is so simple that it could enable the "mass production of these hybrid insect-machine systems," the scientists wrote.

Still, the robo-bugs aren't *quite* ready for their tour of duty. Our directional control is still pretty crude. Ultimately, we'll want to do more than make an insect simply veer left. We'll want to be able to command it to turn, say, precisely 35 degrees to the left or navigate a complicated three-dimensional space, such as a chimney or pipe. There's also the matter of the surveillance equipment. So far, the main focus has been on building insects that we can steer, but for these cyborgs to be useful, we'll need to outfit them with various sensors and make sure that they can successfully collect and transmit environmental information. And though the cyborg insects power their own flight—something that completely robotic fliers cannot do—the surveillance equipment will need to get its electricity from somewhere.

One intriguing possibility is to use the insect's own wings as a source of power. In 2011, a team of researchers from the University of Michigan announced that they had accomplished just that by building miniature generators out of ceramic and brass. Each tiny generator was a flattened spiral—imagine the head of a thumbtack, if it were shaped from a tight coil of metal rather than a single flat sheet—measuring 0.2 inches across. When they were mounted on the beetle's thorax, these generators transformed the insect's wing vibrations into electrical energy. With some refinement, the researchers note, these energy-harvesting devices could be used to power the equipment toted around by cyborg bugs.

Insects could give us a cyborg-animal air force, zooming around the skies and searching for signs of danger. But for terrestrial missions, for our cyborg-animal army, we'd have to look elsewhere.

We'd have to look to a lab at the State University of New York (SUNY) Downstate, where researchers have built a remote-controlled rat.

We've been rooting around in rat brains for ages; neuroscientists often send electrical signals directly into rodents' skulls to elicit certain reactions and behaviors. Usually, however, this work requires hooking a rodent up to a system of cables, severely restricting its movement. When the SUNY team, led by the neuroscientist John Chapin, began their work more than a decade ago, they wanted to create something different—a method for delivering these electrical pulses wirelessly. They hoped that such a system would free researchers (and rats) from a cumbersome experimental setup, and enable all sorts of new scientific feats. A wireless system would allow scientists to manipulate a rat's movements and behaviors while it was roaming freely and give us a robo-rodent suitable for all sorts of special operations. Rats have an excellent sense of smell, so cyborg rats could be trained to detect the scent of explosives, for instance, and then steered to a field suspected to contain land mines. (The task would pose no danger to the animals, which are too light to set off mines.) Or they could be directed into collapsed buildings and tasked with sniffing out humans trapped beneath the rubble, performing a job similar to the one Maharbiz imagines for his cyborg insects. "They could fit through crawl spaces that a bloodhound never could," says Linda Hermer-Vazquez, a neuroscientist who was part of the SUNY team at the time.

But before any of that could happen, the SUNY scientists had to figure out how to build this kind of robo-rat. They began by opening up a rat's skull and implanting steel wires in its brain. The wires ran from the brain out through a large hole in the skull, and into a backpack harnessed to the rodent. ("Backpack" seems to be a favorite euphemism among the cyborg-animal crowd.) This rat pack, as it were, contained a suite of electronics, including a microprocessor and a receiver capable of picking up distant signals. Chapin or one of his colleagues could sit five hundred yards away from the rat and use a laptop to transmit a message to the receiver, which relayed the signal to the microprocessor, which sent an electric charge down the wires and into the rat's brain.

To direct the animal's movements, the scientists implanted electrodes in the somatosensory cortex, the brain region that processes touch sensations. Zapping one area of the cortex made the rat feel as though the left side of its face was being touched. Stimulating a different part of the cortex produced the same phantom feeling on the right side of the rat's face. The goal was to teach the rodent to turn in the opposite direction of the sensation. (Though that seems counterintuitive, it actually works with the rat's natural instincts. To a rodent, a sensation on the right side of the face indicates the presence of an obstacle and prompts the animal to scurry away from it.)

During the training process, the SUNY scientists used an unconventional system of reinforcement. When the rat turned in the correct direction, the researchers used a third wire to send an electrical pulse into what's known as the medial forebrain bundle (MFB), a region of the brain involved in processing pleasure. Studies in humans and other animals have shown that direct activation of the MFB just plain feels good. (When the scientists gave the rats the chance to stimulate their own MFBs by pressing down on a lever, the animals did so furiously—hitting the lever as many as two hundred times in twenty minutes.) So sending a jolt of electricity zinging down to a rat's MFB acted as a virtual reward for good behavior. Over the course of ten sessions, the robo-rats learned to respond to the cues and rewards being piped into their brains. Scientists managed to direct the rodents through a challenging obstacle course, coaxing them to climb a ladder, traverse a narrow plank, scramble down a flight of stairs, squirm through a hoop, and then navigate their way down a steep ramp.

As a final demonstration, the researchers simulated the kind of search-and-rescue task a robo-rat might be asked to perform in the real world. They rubbed tissues against their forearms and taught the rodents to identify this human odor. They constructed a small Plexiglas arena, filled it with a thick layer of sawdust, and buried human-scented tissues inside. When they released the robo-rats into the arena, the animals tracked down the tissues in less than a minute. The scientists also discovered that the rats that received MFB rewards found the target odors faster and dug for them more energetically than rodents that had been trained with conventional food rewards. As Hermer-Vazquez recalls: "The robo-rats were incredibly motivated and very accurate."

Whether it's rescue rat-bots or bomb-sniffing beetle drones, electronics are helping us create new beasts of burden, allowing us to conscript creatures into the modern animal workforce. These are no mere donkeys, poked and prodded into carrying our bags up steep hills; these animals' brains are being taken hostage, their nervous systems forced to cooperate with our plans. As Maharbiz wrote in an account of his research, "[W]e wanted to be sure we could deliver signals directly into the insect's own neuromuscular circuitry, so that even if the insect attempted to do something else, we could provide a countercommand. Any insect that could ignore our commands would make for a crummy robot."

Is it wrong to take the reins of another creature's nervous system? It certainly *feels* wrong. When we dictate the movements of sentient beings, we turn them into mere machines, no different than those remote-controlled airplanes Maharbiz was trying to emulate. Many animal liberationists and philosophers have argued that one of our obligations to animals is "noninterference"—that animals have the right to be the leaders of their own lives and that we have a duty to leave them alone. Cyborg animals represent an extreme violation of that responsibility. And unlike in wildlife tracking projects, in which our meddling may help save species, deploying cyborg insects and rodents on the battlefield isn't going to do much to benefit animals.

The trouble is that we have to balance this intrusion into the life of another living being against the good that animal-machine mash-ups could do. It's possible to care about animals and want to spare them needless suffering, and yet also decide that sometimes human welfare (say, the life of an American soldier) comes first. In fact, most Americans take this view, according to the psychologist Harold Herzog, who specializes in untangling our relationships with other species. After all, if you insist that an animal's life is worth exactly the same as a human one, no matter what, Herzog says, "you can end up at untenable places." (Such as deciding that you should flip a coin to decide whether to save a puppy or a child from a burning building.) Herzog has found that our attitudes toward other species are nuanced, complicated, and often inconsistent. It's not unusual, he says, to wish we could do without animal experimentation but still be grateful for the lifesaving drugs and treatments such research has made possible. It's not strange to wish scientists would stop squirting shampoo into rabbits' eyes and simultaneously want them to use as many bunnies as they need to find a cure for cancer.

Unless we rule out all use of animals for human purposes, we have to evaluate each application on a case-by-case basis, weighing pain against gain. In the case of the robo-beasts, the animals are anesthetized when the electronics are implanted in their bodies, but recovering from surgery isn't painless. The devices themselves may cause stress, and being piloted around a lab by an ambitious postdoc can't be any great picnic. But the price that animals have to pay for this research is relatively small. (Maharbiz notes that his beetles had normal lifespans—which, in insects, is a none-too-impressive several months—and "flew, ate, and mated just like regular beetles.") Remotely guided rats aren't exactly a cure for cancer, but if they can hunt down mines or find earthquake victims trapped in rubble, they could certainly save human lives. So while the cyborg research can seem creepy, I'm glad that there are scientists out there who are doing it.

The details matter, however. I wouldn't be so keen on the research if the cost to animals were greater—if, say, each electric jolt we sent to an animal's brain caused excruciating pain. Nor would I want to see robo-rats used to string lights along the branches of Christmas trees—an actual suggestion the SUNY researchers made in their patent application.* There's a species effect in play, too. I have no special affection for insects or rodents, and I'd find it a lot harder to sanction the creation of robo-dogs or robo-bonobos.† Maharbiz has noticed this inconsistency, too, though it's not clear where that leaves us, ethically speaking. "Where do you draw the line?" he wonders. "Is there a Disney effect: 'Anything cuter than bunnies I will not neuro-control'?" Or should we base our judgments of cyborg projects on something else? Should we make an ethical distinction between forcing muscles to contract (as Maharbiz's wing electrodes do) and simply rewarding an animal for moving the way we want (as Hermer's brain electrodes do)? Or is it how we *use* the cyborgs that matters?

For his part, Maharbiz says he's motivated more by the chal-

* To be fair, it was tenth on their list of ten possible uses for a remotely guided rodent.

† That's not unusual—our attitudes toward different species are hugely inconsistent. (It's right there in the title of Herzog's book: *Some We Love, Some We Hate, Some We Eat.*)

lenge of seeing what he can make insects do than by imagining how his work will ultimately be used. "Maybe I'm an example of a horrible amoral scientist," he says, "but I think it would be fabulous to show, for example, that I could get a beetle to do a barrel roll, which it would never do in nature." Everyone's ethical barometer is set differently, and we won't all welcome the notion of a barrel-rolling beetle. That's fine with Maharbiz, who notes that most of us haven't sat down and thought through what it means to take over an animal's body, to physically force their muscles and minds to do our bidding. Why would we? Until recently, the idea seemed like pure science fiction. One of the ways his work can be useful, Maharbiz says, is "to get people to think about whether this is something we want to do."

Our options for mind manipulation are expanding as well. While Maharbiz and others are using electrodes and wires to physically force neurons to fire, some geneticists and neuroscientists are developing an alternative approach, engineering animals whose brains can be controlled with flashes of light. The technique, which comes from the hot, young field of optogenetics, relies on opsins, a class of light-sensitive molecules that bacteria, fungi, and plants use to sense sunlight and convert it into energy. In 2005, scientists discovered that they could put opsin genes into mammalian brain cells using an unlikely assistant: a virus. Viruses are experts at delivering DNA; whenever they infect a cell, they dump their own genomes inside. In the early days of genetic engineering, biologists realized that they could get viruses to carry other genes into cells, too. In optogenetics, scientists insert an opsin gene into a virus, then inject the modified virus into the brain of a mouse. The virus infects the neurons, depositing the opsin DNA inside.

The mouse's neurons begin to manufacture their own opsins

and install them in their membranes, the thin, fatty layer that surrounds each cell. In the membrane, the opsins operate as light-sensitive channels; when scientists shine a light on the mouse's brain, the opsin channels open and electrically charged particles rush into the cell. The influx changes the voltage inside the neuron. Different opsins respond to light in different ways—some usher positively charged particles into a neuron, making it more likely to fire. Others admit negatively charged particles, which suppress neural activity.* By attaching a little snippet of regulatory DNA to the front of the opsin gene, researchers can make sure that only certain kinds of neurons produce the light-sensitive molecule. As a result, they can engineer a mouse's brain so that one type of neuron, in one brain circuit or region, responds to a flash of light, while its neighbor is unaffected.

Equipped with this technology, we can make mice do the darnedest things. By turning certain neurons on and off, we can make rodents suddenly fall asleep or awaken. Or we can use a beam of light to activate a set of neurons involved in aggression, turning an otherwise calm mouse into a prizefighter who indiscriminately attacks other rodents—or even inanimate objects. These kinds of experiments hold huge promise for basic research; toggling a neural circuit on and off helps scientists puzzle out how those neurons affect behavior.

In 2011, Edward Boyden, a neuroscientist at MIT, used the tools of optogenetics to wirelessly direct mouse movements. Boyden's team began with mice that had been modified to express opsins in certain neurons in the motor cortex; when exposed to light, these motor neurons would begin to fire. Then they constructed a mouse

* They also respond to different wavelengths of light. Add one kind of opsin to a mouse neuron and it will fire whenever you bathe it in blue light. Using a different opsin lets you silence the neuron with yellow light.

helmet—a headpiece that contained a radio antenna and an array of light-emitting diodes—and mounted it on one of their specially engineered mice. The scientists then sat back and used their wireless transmitter to flick the helmet's lights on and off. When they turned all the lights on, a mouse that had been sitting calmly in its cage immediately began running around. ("It's sort of turning up the volume knob of movement," Boyden reports.) They also discovered that when they illuminated just one side of the helmet, a mouse would start spinning in that direction. (Unlike other optogenetics methods, the helmet is entirely noninvasive; the lights can activate neurons from the outside of the skull.)

Optogenetics gives us another way to bend animals to our will, but Boyden has no interest in using his wireless helmets to create a remote-controlled rodent army. To Boyden, the headset is an important breakthrough because it will expand the kinds of experiments that optogenetics researchers can do and pave the way for novel therapeutic devices. Many scientists in the field imagine implanting optical "prosthetics" in the human brain to treat neurological disorders with light. They dream of being able to selectively activate or deactivate neurons involved in Parkinson's, epilepsy, sleep disorders, addiction, and more. Setting animal brains ablaze is the first step toward that goal.

Even as scientists come up with fancy new methods for commandeering animal brains, Greg Gage and Tim Marzullo, a pair of former neuroscience postdocs, are taking these techniques and making them available to anyone with an Internet connection and a hundred dollars to spare. As graduate students at the University of Michigan, the friends volunteered at local public schools, teaching students about human and animal brains. They were frustrated by the high barrier of entry to neuroscience, finding it odd that while anyone can pick up a telescope and look at the Moon, only advanced college students get the opportunity to see a neuron fire.

In 2009, Gage and Marzullo established Backyard Brains, a company that sells low-cost kits that will turn any interested amateur into a neuroscientist, if only for a day or two. (The company's motto, emblazoned on its custom-made circuit boards, is "Neuroscience for Everyone!") Their first product was a little contraption known as the SpikerBox. On sale for $99.98, the device lets customers observe neural firing in a cockroach in real time. (A set of three roaches is $12 extra.) The procedure is simple: Just insert two needlelike electrodes into a cockroach's leg, and the SpikerBox will do the rest, amplifying the electrical activity of the insect's neurons and transmitting it to an attached computer or smartphone as that characteristic visual pattern of peaks and valleys. The SpikerBox put Backyard Brains on the map, and instructors in thirty-five high schools and a hundred universities have used the kits with their students.*

For their second product, Gage and Marzullo decided to push the boundaries even further, to venture beyond brain observation and into brain control. Taking inspiration from the world of cyborg animals, they created a kit that provides their customers with all the tools they need to take over the nervous system of a living cockroach. In principle, the Backyard Brains RoboRoach is nearly indistinguishable from the beetles Maharbiz is making in a university lab—and that is precisely what is so remarkable about it. It means we can all experiment with bionic bugs in our own homes. Or, as it happens, in a crowded neighborhood coffee shop, which is precisely

* The Backyard Brains website keeps a running tally of how many people have eavesdropped on neural activity for the first time, thanks to the SpikerBox. As of June 2012, the total was 15,809 people and counting.

the plan when I meet Gage and Marzullo for breakfast in Woods Hole, Massachusetts.

The bespectacled pair greet me at a popular local café, and we find ourselves some seats on the outdoor patio. Marzullo pulls out a plastic box of cockroaches and plops it down on our table. If you're new to the hobby of animal mind manipulation, the cockroach is an excellent place to start. Because a roach relies on its long, fluid-filled antennae for a host of sensory and navigational functions, its nervous system is stunningly easy to hack; all a wannabe roach-master has to do is thread a wire inside each antenna. ("It's like *designed* to be a cyborg," Marzullo says.)

Marzullo has spent the morning prepping two roaches for their remote-controlled destiny. Several hours ago, he dropped the cockroaches into a miniature cooler of ice water—the preferred method, apparently, for anesthetizing insects. Then he pulled the roaches out of the cooler, their bodies motionless, their sensations dulled. ("We actually don't know if insects feel pain," Marzullo and Gage write on the Backyard Brains website, "but we do make the assumption that they do, which is why we anesthetize them in thc first place.") With a pair of everyday household scissors, Marzullo snipped the ends off each antenna. Then he slipped a thin silver wire inside. Thereafter, any electrical signals sent down the wires would be transferred directly to the roach's nervous system.

Steering the roach simply requires taking advantage of a natural cockroach instinct: When one of the cockroach's antennae detects an obstacle, the bug turns in the other direction. Zap the right antenna and the insect, convinced it's about to bump into a wall on the right side of its body, will turn to the left. And vice versa. (The SUNY researchers had tapped into the same instinct in training their cyborg rats to turn away from perceived obstacles. But unlike the robo-rats, the cyborg cockroaches needed no special training or reinforcement to follow directional commands.)

Marzullo opens his bug box and removes one of the roaches. The wires run out of its antennae and into a small black box that Marzullo has glued onto its head. Marzullo plugs this "connector" into the cockroach backpack, a red-and-green assemblage of circuit boards. The electronics are slightly modified versions of circuit boards that come from a widely available toy: a plastic, remote-controlled inchworm called the HexBug that retails for twelve dollars at Toys"R"Us. When these circuit boards are linked to the head-mounted connector, Marzullo and Gage can use the remote control that comes with the toy to deliver pulses of electricity to the roach.

As Marzullo fiddles with the cockroach, he notices a family of three sitting at the table next to us. They're all staring.

"What is it?" the father asks.

"The world's first commercially available cyborg," Marzullo says. "You want to do it, young lady?" he asks, handing the remote to the man's ten-year-old daughter. He shows the girl which buttons to press.

We all head out to the sidewalk. The bug goes down. The little girl starts hitting buttons on the remote, steering the roach all around the sidewalk, while her father advises: "Don't let it go into the street . . . Turn him into the shade." The girl's power to control the roach is, admittedly, crude. She can't make the insect start or stop moving, and there's no way to force it to simply move forward in a straight line. All she can do is let the roach do its roach thing, taking off in whatever direction its little invertebrate heart desires, and then overlay her own "left" or "right" commands, forcing the bug to turn and start moving in a different direction.

But even that small power is impressive, and a crowd forms.

People watch and smile, and Gage and Marzullo laugh and joke with the assembled audience. "There you go," Marzullo says, "neuroscience for the people."

"It looks so real!" a passing woman exclaims.

"It *is* real," Gage says. "We're selling these for ninety-nine bucks." The kit comes with all that customers need to make the cyborgs themselves—the circuit boards, the controller, the remote, and detailed instructions for performing the insect surgery.

Then it's my turn. I return to our table and pick up the second roach, which Marzullo has kindly prepared for me. Its sticky legs tickle my palm as I carry it out to the sidewalk. I place it down gingerly, and it begins to scuttle off. I fumble with the remote before finding the "L" button. I hit it, and the roach abruptly spins to the left. The effect is less dramatic after that, but convincing.

"It's such a compelling demonstration," Marzullo says. "We go to classrooms all the time and even the most jaded, problem kid in the room is going to pay attention to this. It doesn't take very much time to break down their veneer when we bring out . . . remote-controlled bugs."

Nevertheless, the RoboRoach has not been nearly as brisk a seller as Gage and Marzullo had hoped; as of June 2012, they'd squeaked out fifty-one sales. Perhaps that's because it takes a special kind of customer to want to hijack another creature's mind. "It's kind of edgy," Marzullo says. "It taps into human fears of puppet masters, that we are somehow evil scientists that don't respect the natural order of things." Gage and Marzullo have heard the same objections as other cyborg-animal scientists—that what they're doing to animals is inhumane, disgusting, and just plain wrong. They say they inspire more vitriol than scientists like Maharbiz, who are doing their research in official university laboratories. "We're doing all this stuff on the fringe," Marzullo says. "We're not affiliated with any university, and we go out in public, and we're pretty flamboyant about what we do."*

Gage and Marzullo attract controversy for the same reason that Alan Blake did as he prepared to bring GloFish to market—because they are taking biotechnology out of the lab and putting it into the hands of the public. And just like Blake, they are criticized for meddling with animal bodies for "trivial" purposes. Most people, Marzullo explains, have accepted the use of animals for scientific research, defense, or food. "But if you exploit animals for education," he says, "people aren't cool with that." ("From my perspective," he adds, "that's the best use of animals. It's an investment in the future.")†

Is educating students about the nervous system—and potentially encouraging a new generation of neuroscientists—a less-justifiable use of animals than hunting out mines or earthquake survivors? It's time to start thinking through these issues, because now that the tools of brain control have been liberated from the lab, there's no telling how they'll be used.

Indeed, there is a growing community of "biohackers," science enthusiasts who are experimenting with genes, brains, and bodies outside the confines of traditional laboratories, working on shoestring budgets in their garages and attics, or joining the community labs that are springing up around the country. Some of these resourceful do-it-yourselfers are even building their own versions of

* For the most part, invertebrates (such as cockroaches) are not protected by federal or institutional regulations on animal research and experimentation. It's not clear whether or how the government would respond if Gage and Marzullo decided to sell, say, robo-rat kits, but any use of vertebrates, such as rodents or birds, would likely open them up to more legal scrutiny.

† The pair rarely lets a teaching opportunity pass them by. While on a flight together, they once posted a sign on the airplane bathroom door advertising FREE NEUROSCIENCE LESSONS at seats 33A and 33B.

high-tech laboratory equipment that normally costs thousands of dollars.

Backyard Brains is tapping into this movement, giving amateurs access to some of science's most sophisticated tools and techniques. (As it happens, their most recent product is a kit that allows customers to play around in the world of optogenetics, using blue light to make the muscles of transgenic fruit flies twitch.) And their customers are surprising them, in the best possible way, by coming up with ideas and discoveries of their own. A class of New York high school students working with the RoboRoach pinpointed a nerve that they could stimulate to make the insect walk straight ahead. Another customer—a Microsoft programmer—bought an EEG cap and tried to use his own brain waves to steer the roach. (It didn't work, but he gets points for creativity.)

If these unprompted experiments are any indication, there are plenty of amateurs with an appetite for independent investigation and their own ideas for reengineering animals. Future generations are going to grow up tinkering not with computers, but with life itself. We already have the annual International Genetically Engineered Machine competition, in which high school and college students use standard genetic parts—easily available bits of DNA—to create cells with novel properties. In past years, teams have created bacteria that can clean heavy metal from polluted water, glow in a rainbow of colors, or give off the pleasant odor of banana or mint. We may one day have a similar competition that asks youngsters to engineer new kinds of animal-machine hybrids. Perhaps DARPA will even invite enthusiastic amateurs to respond to its scientific calls or look to the public for solutions to its most pressing problems.

The latest, greatest cyborg critters may come not from state-of-the-art labs, but the minds of curious kids and individual hobbyists. Though scientists will continue to build their cyborg animals, Maharbiz says he fully expects that "kids will be able to hack these things, like they wrote code in the Commodore 64 days." We are heading toward a world in which anyone with a little time, money, and imagination can commandeer an animal's brain. That's as good a reason as any to start thinking about where we'd draw our ethical lines. The animal cyborgs are here, and we'll each have to decide whether we want a turn at the controls.

8. Beauty in the Beasts

Over the last few decades, animals have gained status in the Western world. We increasingly treat other species as worthy of moral consideration, and livestock, pets, and research animals enjoy far more protections than they once did. For instance, there is a movement under way to grant great apes legal (or even human) rights, and many governments have severely curtailed research on primates. In December 2011, the National Institutes of Health announced that it was indefinitely suspending the funding of new research on chimpanzees, pending further review, and a bill currently before Congress would outlaw invasive research on all great apes. A number of other nations—including the United Kingdom, New Zealand, Austria, the Netherlands, Belgium, and Sweden—have similar laws in place, and in 2010, the EU passed a ban on most research on great apes, to take effect in 2013.

We're elevating the standing of pets, too. Some American cities have passed laws that say humans aren't owners of their animal companions—instead, they are guardians of them. Seventy percent of dog owners in the United States now consider their animals to be legitimate members of the family, and Americans spend $48 billion a year on their pets.

However, these figures pale in comparison with the $300 billion we spend every year eating animal flesh. When Harold Herzog, the psychologist who specializes in human-animal relations, surveyed his own students, he discovered that nearly half of them agreed with the statement "Animals are just like people in all important ways." But of these students who put animals and people on the same plane, 90 percent of them ate meat and 50 percent supported xenotransplantation. National surveys have turned up similar findings. In a Gallup poll, 71 percent of Americans said that animals deserved "some protection from harm and exploitation," and an additional 25 percent said that animals should have the same rights as people. Yet 64 percent of all respondents accepted the practice of using animals in medical research. More astoundingly, of those who said that animals deserved the exact same rights as humans, 44 percent supported at least the occasional use of animals in such research.

These conflicting attitudes position most of us in a terrain that Herzog calls the "troubled middle" (a term he credits to the philosopher and bioethicist Strachan Donnelly). The troubled middle is a land of contradictions. It's a place where it's possible to truly love animals and still accept their occasional role as resources, objects, and tools. Those of us in the troubled middle believe that animals deserve to be treated well, but we don't want to ban their use in medical research. We care enough to want livestock to be raised humanely, but don't want to abandon meat-eating altogether. "Some argue that we are fence-sitters, moral wimps," Herzog, himself a resident of the troubled middle, writes. "I believe, however, that the troubled middle makes perfect sense because moral quagmires are inevitable in a species with a huge brain and a big heart. They come with the territory."

Even Charles Darwin was a resident of the troubled middle. Darwin hated animal cruelty, but famously refused to condemn invasive animal research. "I know that physiology cannot possibly progress except by means of experiments on living animals," he wrote, "and I feel the deepest conviction that he who retards the progress of physiology commits a crime against mankind."

For the vast majority of us who reside in the troubled middle, there are no easy answers to the ethical dilemmas that biotechnology can pose. As biotechnology moves forward, we'll have to carefully evaluate each application on its own terms, trying to balance what's in the best interests of an individual animal with what's good for its species as a whole, for humanity, and for the world that we all share. Even if we decide that there are instances in which animal discomfort is justified, we should take this discomfort seriously. We do that by making sure animals' pain is controlled—by anesthetizing them, for instance, before performing surgical procedures—by meeting their physical and psychological needs while they're living in laboratories, and by keeping the numbers of experimental animals as low as possible.

Perhaps the most valuable thing our expanding scientific capabilities will do is spark a real dialogue about our interactions with the other living beings that populate this planet. "We've always had strong moral responsibility, or we should have had, to other species," says Richard Twine, the British sociologist. "We just haven't exercised that very well." Biotechnology provides an opening for reconsidering our obligations to animals. How can we seize this moment to rethink our relationships with other species and recommit to their well-being?

For starters, the techniques of today and tomorrow could help us reverse the pain and suffering we've inflicted upon the animal kingdom. For an example, we need look no farther than the burgeoning field of canine genetics. Over the generations, we have bred and inbred our canine companions to the point of disease and deformity. One analysis of fifty popular dog breeds turned up a total of 396 inherited diseases affecting the canines; each breed included in the analysis had been linked to at least four, and as many as seventy-seven, different hereditary afflictions. Dalmatians are prone to deafness, Dobermans suffer from narcolepsy (the image of a fierce canine suddenly nodding off would be funny if it weren't so pathetic), and Labrador retrievers are renowned for their terrible hips. In some cases, these disorders are nasty side effects of a small gene pool, of generations of breeding related dogs or relying on just a few popular sires. In others, they're due to intentional selection for the exaggerated physical traits prized by kennel clubs and dog show judges.*

Thanks to modern genetics and genomics, we're developing the tools to undo the damage we've done and conquer many canine ailments. As of 2012, commercial labs in North America, Europe, and Australia were offering tests for eighty genetic mutations linked to doggy diseases. For less than a hundred dollars, for example, VetGen can tell you whether your beagle has genetic variants that cause a bleeding disorder or your Boston terrier has a mutation linked to early onset cataracts, vital information that could help you secure the right medical care for your pooch. Breeders have also started using DNA testing results to create healthier dogs in the first place, carefully arranging matings that reduce the number of puppies prone to serious illnesses. Many canine diseases are in-

* For example, Cavalier King Charles spaniels, the breed that gave my Milo half of his DNA, have been bred to have domed heads. The result is a skull that's too small and underdeveloped for the dogs' brains, causing spinal cord problems, brain damage, and chronic pain in many Cavaliers.

herited in a recessive pattern, meaning that a dog has to have two copies of a disease-causing mutation in order to develop the disorder. In these cases, dogs with a single copy of the mutation are known as carriers—they're healthy, but they can pass the bad gene on to their puppies. When two carriers mate, there's a chance that some of the puppies will inherit the unhealthy variant from both parents and will, in turn, get sick. Genetic testing can reveal which dogs are free of recessive, disease-causing mutations and can thus be bred without restriction. The carriers can breed too, as long as they mate with non-carriers. In this way, we can reduce the number of dogs that develop serious ailments while allowing the maximum number of pooches to contribute their genes to future generations. As it happens, genetic testing, followed by thoughtful breeding, has already reduced the number of English springer spaniels carrying a gene for a metabolic disease and the prevalence of progressive blindness among Irish setters and corgis.

Identifying the genes responsible for disease also opens up new possibilities for treatment, including gene therapy, in which vets give their canine patients a "healthy" version of whatever gene has gone haywire. Gene therapy experiments have been remarkably successful in dogs, with one project even giving blind dogs the gift of sight. These dogs had all been blind since birth, due to a mutation in a gene known as RPE65, which normally codes for a protein crucial to vision. In 2001, Gustavo Aguirre, a veterinary ophthalmologist and geneticist at the University of Pennsylvania, and his colleagues engineered a virus that contained a healthy form of RPE65. They injected the virus into the eyes of their blind canine patients. The viruses delivered the new RPE65 gene into the dogs' cells, which then started churning out a fully functional version of the critical protein—for the first time in these animals' lives. Within two weeks, the dogs' vision began to improve; within four months, they were able to successfully navigate a laboratory obstacle course.

And the fix was permanent; the first canine patient lived for eleven years after the gene therapy, able to see until its dying day. (For blind animals—and humans—that aren't good candidates for gene therapy, scientists are working on another option: bionic eyes, or retinal prostheses. The approach involves implanting electrodes in the eye that stimulate the retinal cells.)

Genetic technology isn't just for diseases that have an obvious heritable component. In 2011, Helen Sang, a developmental biologist at the Roslin Institute, and Laurence Tiley, a virologist at the University of Cambridge, engineered transgenic chickens that are incapable of spreading avian influenza to the other members of their feathered flocks. (Alas, the modified chickens can still contract the flu themselves—they simply don't transmit the normally contagious illness to other birds.) The modification could save the lives of countless chickens and reduce the threat to human health, representing the ultimate win-win. In fact, Duane Kraemer, the scientist who helped clone several species, thinks that biotechnology has so much potential to improve the health of farmed animals—and to safeguard the health of humans along the way—that "cloned" and "genetically engineered" may one day acquire the same cachet as "organic" or "free range." "What I would like to see happen," he says, "is for the products and the strains of animals that are developed—for people to become so proud of those that they'll advertise: 'These are cloned and genetically engineered products! And they're special!' I think someday that will happen, and that's when the public will of course be much more accepting."

Of course, even endeavors that seem as straightforward as making disease-resistant animals can be fraught with ethical complexity. In some situations—such as when we're dealing with farm animals—our motivations may not be entirely altruistic. "You have to bear in mind the economic context of their lives," Twine says. "Clearly, the main motivation for making animals disease-resistant is to maximize profitability of their existence as commodities. There could be some benefit in terms of an animal perhaps having less degrees of suffering, but of course they don't escape that category of commodified farmed animal. They still have to face an early death in the slaughterhouse." Furthermore, operators of factory farms may view the creation of healthier and hardier animals as an excuse to cram livestock into crowded pens, let them live in unhygienic conditions, and otherwise treat them poorly.

Or consider a more extreme prospect, laid out in a 2010 *New York Times* editorial headlined NOT GRASS-FED, BUT AT LEAST PAIN-FREE. In it, Adam Shriver, a graduate student at Washington University in St. Louis who specializes in philosophy and neuroscience, outlined a remarkable bit of research. Scientists had discovered, he wrote, how to genetically engineer mice that were missing enzymes critical to the brain's pain-processing system. That made the rodents unable to feel pain, as though they were hooked up to a permanent morphine drip. Shriver set forth a radical proposal: Given the animal suffering inherent in the meat industry, and the fact that humans aren't likely to abandon their carnivorous ways anytime soon, we should start genetically engineering livestock that feel less pain. "If we cannot avoid factory farms altogether," he wrote, "the least we can do is eliminate the unpleasantness of pain in the animals that must live and die on them." Every logical bone in my body agrees with Shriver, yet the emotional part of me resists. Though the ostensible goal of engineering pain-free animals is to minimize other species' discomfort, what it's really doing is alleviating our own. If we think these creatures aren't capable of feeling much pain, will that give us license to alter and exploit them in even more profound ways?

These are the discussions we'll need to have if we have any hope of using our new technologies responsibly. Do we want to make genetically engineered, disease-resistant livestock so that we can get away with substandard living conditions and inadequate medical care, while maximizing profits on factory farms? Or do we want to use these creatures as an opportunity to launch a broader campaign to improve the lives of farm animals? In some ways, our own unease with these technologies is productive—it means we will have to keep evaluating and reevaluating their consequences for animals.

The important thing is that we do not throw the genetically modified baby out with the bathwater. We spend so much time discussing the ethics of using our emerging scientific capabilities that we sometimes forget that *not* using them has ethical implications of its own. How many animals (and humans) will suffer if we turn our backs on breakthroughs like a genetically engineered chicken incapable of spreading the flu? Biotechnology is not the only solution to what ails animals, but it's a weapon we now have in our arsenal, one set of strategies for boosting animal health and welfare. If we reject it out of hand, we lose the good along with the bad.

If we really want to boost animal welfare, perhaps we should be embracing technology, not running from it. That's what George Dvorsky, a Canadian bioethicist and futurist, believes. Dvorsky, who heads up the Rights of Non-human Persons program at the Institute for Ethics and Emerging Technologies, says we owe animals far more than merely leaving them alone. Instead, he thinks we have a responsibility to use all the scientific techniques at our disposal to improve their lives. As a society, he says, we are increasingly toying with the prospect of *human* enhancement, with our growing ability to use some combination of pharmacology, genetics, and electronics to upgrade our bodies and brains. In Dvorsky's mind, if we're going to build a better version of our own species, animals should get the benefit of the same technologies.

One option: enhancing animals' sensory skills. For instance, while dogs have a great sense of smell, their vision isn't quite so spectacular. "Their horizon line is extremely low," Dvorsky says. "They don't see in the broad spectrum of colors that we do." The right genetic manipulation or brain chip might change that. Dvorsky also imagines making dramatic upgrades to animal cognition, altering the genome of a bonobo in ways that supercharge its memory, for instance, or boost its capacity for using complex forms of language. "I realize how absolutely extreme that sounds," he admits. "It's really, really out there. But I'm doing my duty as a thinking person. Just because we lucked out in the genetic lottery doesn't mean that we don't have a moral responsibility and obligation to the other animals of the planet."

Dvorsky's dream of memory enhancement is not as far-fetched as it may seem. Scientists have already engineered dozens of strains of "smart mice," which learn faster and retain more than their non-modified counterparts. Another team of researchers managed to improve rats' performance on a memory test by using implanted electrodes to stimulate neurons in the hippocampus, a brain structure involved in memory formation and storage.

Dvorsky has been criticized for being an interspecies imperialist, for suggesting that animals would be better off if they were more like us. But that's not quite what he's saying; he wants to augment animals' natural talents and capabilities, which may or may not actually make them more humanlike. In fact, Dvorsky says we could improve ourselves by adopting certain *animal* characteristics—the vision of a hawk, for instance, or the ability to swim underwater for extended periods of time, like a dolphin. What he imagines, he says, is a total "blurring of the species line," the scientific elevation of "the entire biosphere"—humans, dolphins, and dogs all gaining new capabilities together.

It's still unclear whether we'll ever be able to enhance the bonobo's language skills and, more pertinently, whether doing so would improve the ape's quality of life. But I agree with Dvorsky that there are instances in which engineering (or reengineering) animals is a moral imperative.

The world is becoming ever more human, increasingly created by us, for us. We dam rivers, till land, and clear forests right under the feet and fins of the creatures that live there. We spray plants with toxic fertilizers and dump our industrial waste in lakes and rivers. We take far-flung vacations, allowing foreign species to make their way into new lands. (In fact, we are changing the environment so completely that geologists have given our epoch a new name: the Anthropocene, from the Greek root *anthropo*, which means "human.") Then there's climate change, which is altering the slim slices of habitat animals have left. Some species will adapt, of course; as the planet warms, birds have expanded their ranges northward. But for others, the rapid pace of change we're causing will simply be too much. The United Nations Intergovernmental Panel on Climate Change estimated that an increase of 3.5 degrees Celsius in global temperature could result in the extinctions of anywhere from 40 to 70 percent of the planet's species.

Even when we're not driving entire species toward extinction, we remain a powerful evolutionary force, capable of transforming the bodies of wild animals. Consider the impact that our hunting and harvesting has had on entire populations. Though a big ram with large antlers is the last animal a wild predator would target, human hunters covet these impressive specimens. We have harvested so many of these large deer, elk, and sheep over the centuries that many species have evolved smaller body and horn sizes. Similarly, fish have adapted to human harvesting by developing thinner bodies capable of sneaking out of nets.

Humans are a force of nature—we are, in some senses, *the* force of nature—and we influence animals whether we intend to or not.

So the real question, going forward, is not *whether* we should shape animals' bodies and lives, but *how* we should do so—with what tools, under what circumstances, and to what end. Are the needs of other species truly best served by leaving them to fend for themselves in a world that we have come to dominate? Unless we plan to move all humanity to Mars and leave Earth to rewild itself, we may need to help our furry and feathered friends survive in a world that has us in it. As Kraemer puts it: "I'm of the persuasion that we are changing the habitat for wildlife so rapidly that we may have to help those species evolve."

We've only scratched the surface of what's possible. We've seen how scientists are already changing animal lives and considered how their work might play out in the near future. But what about the more distant one? If we went on a tour of the world's pet stores, nature preserves, and family farms fifty or a hundred years from now, what would we see? There are enough journalists, politicians, and ethicists out there speculating about the worst-case scenarios—the glowing teenagers, the resurrected Hitlers, the killer cyborg armies. They've got the apocalyptic visions covered. After my time in the land of cloned creatures and bionic beasts, I'm ready to imagine an alternative future, one in which biotech brings hope and promise rather than anxiety and alarm.

In this world, I envision stocking our fields and farms with healthier animals. We'll find cows, goats, and horses that are naturally resistant to disease and then clone them. When we can't find such mutants, we'll make them, engineering livestock that are free from diseases that threaten both humans and animals. We could modify all cows so that their milk contains higher levels of antibacterial compounds or heart-healthy fats. That way, we wouldn't need a special prescription for supermilk—it's what we'd all be drinking by default. And we could create critters whose milk is better for their *own* nursing offspring.

We could also equip all farm animals with tiny electronic devices, such as the temperature-sensing microchips that are beginning to make their way onto the market. When injected just underneath the skin, these "Bio-Thermo" chips continuously monitor a critter's internal body temperature. Imagine deploying these devices on a massive scale, putting one in every farm animal as soon as it's born. If the world's farms all contained microchipped cows, goats, pigs, and chickens, we could monitor the animals for sudden signs of fever and use the temperature spikes as an early warning sign of a possible disease outbreak.

Maybe we could engineer these chips to measure other health indicators as well—blood pressure, hormone levels, and more—and combine them with wildlife tags. The tracking devices of the future could tell us not just *where* animals are, but also *how* they are. Are elephant seals thriving? Or are they just getting by? We could tag a large and representative sample of the seal population, keeping our eyes peeled for an unexpected rise in the rates of illness or death. The data might be able to help us identify an impending population catastrophe and give us the opportunity to intervene before it's too late.

Closer to home, we could put these kinds of chips in our pets as well. My dream? Networking these devices with our smartphones. Envision the kinds of apps we could create: An alert pops up on your phone. It tells you that Fido's got a fever, but that none of his other vital signs seem out of whack. Given that the fever's not high, the program recommends watchful waiting, but notes that if the dog develops serious vomiting or diarrhea, you may want to take him to the vet. The software provides a list of links for you to explore if you'd like to read more about the possible causes of canine fever and reassures you that it will ping you with updates every hour until Fido's condition improves. I'd pay a lot for a system like that (and can think of more than one panicked call to the emergency animal hotline that it might have prevented).

When it comes to dog diseases, future pet owners may be better equipped to manage the occasional defects and abnormalities that do pop up. What if every new puppy came with a readout of its complete genetic profile? Armed with this information, we'd be able to provide our dogs with better medical care, monitoring them for early signs of illness and formulating treatment plans that keep them healthy as long as possible. We may be able to nip all sorts of problems in the bud with an early dose of gene therapy. Better yet, we might be able to fix defective genes in dog eggs, sperm, and embryos. That would not only keep individual dogs from suffering, but also make more dogs eligible for breeding, thereby keeping the canine gene pool as diverse as possible. (Not a dog lover? Never fear. DNA tests and screening programs for cats and horses are beginning to proliferate, too.)

We may be able to harness other laboratory breakthroughs to nudge all the world's animals one step closer to immortality. One potential tweak involves a gene that codes for a metabolic enzyme that goes by the nickname "PEPCK-C." (It doesn't exactly roll off the tongue, but it's much better than the enzyme's full name: phosphoenolpyruvate carboxykinase.) PEPCK-C helps our bodies produce the glucose that our cells use as fuel. In 2006, scientists at Case Western Reserve University engineered mice that made elevated levels of PEPCK-C in their muscles. This single alteration had far-reaching effects. For one, the modified rodents were natural marathoners, capable of running for hours at a time without stopping. Normal mice tired and quit after just 0.2 kilometers on a mouse treadmill; the modified mice went twenty-five times as far, cranking out 5 kilometers at a stretch. Even more remarkably, the engineered mice lived two years longer than normal mice, and the females were fertile for twice as long. What if we made this same genetic modification in endangered species? It would give us animals that not only lived longer, but also had more opportunities to reproduce in the wild. This one small genetic change could be enough to help threatened populations rebound.

My crystal ball of biotechnology reveals other ways we could help animals transcend their biological limits. Wouldn't it be wonderful if instead of euthanizing every broken-down racehorse, we simply gave them all bionic legs? (Of course, it would be even better if we stopped racing horses altogether, but barring that, prosthetics may at least provide a way to keep more of these equines alive after catastrophic injuries.) Or we could push the field of animal prosthetics even further: What if we replaced the legs of aging dogs with springy prostheses that let them run faster and farther than they ever did as puppies? Or gave the future Winters of the world motorized tails, boosting the cetaceans' speed and enabling them to flip and spin and perform exciting new tricks? We could make injured or elderly animals not just functional again, but better than nature ever intended. Bionic limbs might help our beloved creature companions stay spry as they age and squeeze as much life as they could out of each of their days.

These ideas might seem like pie-in-the-sky fantasies, but imagining a future in which we elevate animals and enhance their lives is the first step in bringing that world into being. And it's not just animals that stand to gain. Indeed, we've already seen how technology can jump across species barriers. The prosthetic liner designed for a cheeky dolphin ended up solving major problems for human amputees. Some of the vision disorders that affect dogs have close analogues in humans, and the gene therapy that cured dogs of their blindness is being tested in visually impaired people. Optogenetics also promises revolutionary new treatments for human neurological disorders. As science advances, I suspect we'll see more and more of this kind of crossover, with innovations in the animal world inspiring breakthroughs in the human one (and vice versa). In 2012, for instance, a team of Swiss researchers used chemical infusions and implanted electrodes to stimulate the spinal cords of paralyzed rats. The treatment helped the rodents get back up and running again—literally—and it may one day do the same for paralyzed humans. By enhancing animals, we may discover ways to make *ourselves* smarter and stronger, faster and fitter, healthier and happier.

Biotechnology is not inherently good or bad; it is simply a set of techniques, and we have choices about how we employ them. If we use our scientific superpowers wisely, we can make life better for all living beings—for species that walk and those that fly, slither, scurry, and swim; for the creatures that live in scientific labs and those who run them. So it's time to embrace our role as the dominant force in shaping the planet's future, time to discover what it truly means to be stewards. Then we can all evolve together.

Notes

Introduction

3 *a new industry is taking shape*: The researcher leading this effort, Tian Xu, has a dual appointment at Fudan University and Yale University, but the mouse manufacturing itself is happening at Fudan. Unfortunately, Xu did not respond to repeated requests for interviews, so information about his work and the technique he developed comes from several sources, including S. Ding et al., "Efficient Transposition of the PiggyBac (PB) Transposon in Mammalian Cells and Mice," *Cell* 122 (2005): 473–83. Ling V. Sun et al., "PBmice: An Integrated Database System of PiggyBac (PB) Insertional Mutations and Their Characterizations in Mice," *Nucleic Acids Research* 36 (2008): D729–34. Sean F. Landrette and Tian Xu, "Somatic Genetics Empowers the Mouse for Modeling and Interrogating Developmental and Disease Processes," *PLoS Genetics* 7, no. 7 (2011). Sean F. Landrette et al., "PiggyBac Transposon Somatic Mutagenesis with an Activated Reporter and Tracker (PB-SMART) for Genetic Screens in Mice," *PLoS One* 6, no. 10 (2011). Muyun Chen and Rener Xu, "Motor Coordination Deficits in *Alpk1* Mutant Mice with the Inserted *PiggyBac* Transposon," *BMC Neuroscience* 12, no. 1 (2011). "Pioneering New Genetic Tools & Approaches," Tian Xu Laboratory, accessed March 5, 2012, http://info.med.yale.edu/genetics/xu/index.php?option=com_content&task=view&id=37. "Deciphering Mammalian Biology and Disease," Tian Xu Laboratory, accessed June 1, 2012, http://info.med.yale.edu/genetics/xu/index.php?option=com_content&task=view&id=33. "PBmice: Piggybac Mutagenesis Information Center," Fudan University," accessed June 1, 2012, http://idm.fudan.edu.cn/PBmice/. "Tian Xu, Ph.D." Howard Hughes Medical Institute, accessed March 23,

2012, www.hhmi.org/research/investigators/xu_bio.html. Dennis Normile, "China Takes Aim at Comprehensive Mouse Knockout Program," *Science* 312 (June 30, 2006): 1864. Pat McCaffrey, "Little Mouse, Big Medicine," *Yale Medicine*, Winter 2007, http://yalemedicine.yale.edu/winter2007/features/feature/51773. Margot Sanger-Katz, "Building a Better Mouse," *Yale Alumni Magazine*, May/June 2010, www.yalealumnimagazine.com/issues/2010_05/mouse349.html. Michael Wines, "China Lures Back Xu Tian to Decode Mouse Genome," *New York Times*, accessed June 1, 2012, www.nytimes.com/2011/01/29/world/asia/29china.html?pagewanted=all.

5 *Exactly how this dog domestication began*: Various accounts of dog domestication are available in Adam Miklosi, *Dog Behaviour, Evolution and Cognition* (Oxford, UK: Oxford University Press, 2007), and James Serpell, ed., *The Domestic Dog: Its Evolution, Behaviour, and Interactions with People* (Cambridge, UK: Cambridge University Press, 1995). For an accessible exploration of the theory that dogs helped to domesticate themselves, see Stephen Budiansky, *The Covenant of the Wild* (New Haven, CT: Yale University Press, 1999).

5 *Their bodies and heads shrank*: Information on how and why wolves' bodies changed comes from Miklosi, *Dog Behaviour, Evolution and Cognition*; Serpell, *The Domestic Dog*; Susanne Bjornerfeldt et al., "Relaxation of Selective Constraint on Dog Mitochondrial DNA Following Domestication," *Genome Research* 16 (2006): 990–94; and Helen M. Leach, "Human Domestication Reconsidered," *Current Anthropology* 44, no. 3 (2003): 349–68.

5 *We created the . . . badger burrows*: "Mastiff," American Kennel Club, accessed March 5, 2012, www.akc.org/breeds/mastiff/. "Dachshund," American Kennel Club, accessed March 5, 2012, www.akc.org/breeds/dachshund/.

5 *One year, the "Best in Show"*: These dogs were finalists in 2009.

5 *most physically diverse species*: Taryn Roberts et al., "Human Induced Rotation and Reorganization of the Brain of Domestic Dogs," *PLoS One* 5, no. 7 (2010).

6 *We've reshaped other species*: Jared Diamond, "Evolution, Consequences and Future of Plant and Animal Domestication," *Nature* 418 (August 8, 2002): 700–707.

7 *consider a very curious cat*: Martha C. Gomez, "Generation of Domestic Transgenic Cloned Kittens Using Lentivirus Vectors," *Cloning and Stem Cells* 11, no. 1 (2009): 167–75.

7 *a barn in Logan, Utah*: "Synthetic Silk," Utah State University, accessed March 5, 2012, http://sbc.usu.edu/htm/silk; Adam Rutherford, "Synthetic Biology and the Rise of the 'Spider-Goats,'" *The Guardian*, January 12, 2012, www.guardian.co.uk/science/2012/jan/14/synthetic-biology-spider-goat-genetics; Geoffrey Fattah, "USU Goats May Be Key to One of the Strongest Known Substances," July 10, 2011, www.ksl.com/?nid=960&sid=16249521. "The Goats with Spider Genes and Silk in Their Milk," BBC News, January 16, 2012, www.bbc.co.uk/news/science-environment-16554357.

7 *to use tiny electrodes*: S. K. Talwar et al., "Rat Navigation Guided by Remote Control," *Nature* 417, no. 6884 (May 2, 2002): 37–38. S. Xu et al., "A Multichannel Telemetry System for Brain Microstimulation in Freely Roaming Animals," *Journal of Neuroscience Methods* 133, no. 1–2 (2004): 57–63.

7 *Breakthroughs in materials science*: Noel Fitzpatrick et al., "Intraosseous Transcutaneous Amputation Prosthesis (ITAP) for Limb Salvage in 4 Dogs," *Veterinary Surgery* 40, no. 8 (2011): 909–25.

7 *we can train monkeys*: M. Velliste et al., "Cortical Control of a Prosthetic Arm for Self-feeding," *Nature* 453 (June 19, 2008): 1098–1101; Jose M. Carmena et al., "Learning to Control a Brain-Machine Interface for Reaching and Grasping by Primates," *PLoS Biology* 1, no. 2 (2003).

8 *we've saddled dog breeds*: L. Asher et al., "Inherited Defects in Pedigree Dogs. Part 1: Disorders Related to Breed Standards," *The Veterinary Journal* 182 (2009): 402–11; J. Summers et al., "Inherited Defects in Pedigree Dogs. Part 2: Disorders That Are Not Related to Breed Standards," *The Veterinary Journal* 183 (2010): 39–45.

8 *turkeys with such gigantic breasts*: Alan W. Dove, "Clone on the Range: What Animal Biotech Is Bringing to the Table," *Nature Biotechnology* 23 (2005): 283–85.

10 *60 percent of Americans*: Frank Newport et al., "Americans and Their Pets," Gallup News Service, December 21, 2006, www.gallup.com/poll/25969/americans-their-pets.aspx.

1. Go Fish

14 *But these swimmers are adulterated . . . first genetically engineered pets*: Alan Blake, in discussion with author, via telephone, January 14, 2010; Alan Blake, e-mail message to author, February 1, 2012; "GloFish Fluorescent Fish," Yorktown Technologies, accessed February 1, 2012, www.glofish.com/; Alan Blake, "GloFish—The First Commercially Available Biotech Animal," *Aquaculture Magazine*, November/December 2005, 17–26.

15 *They isolated small stretches*: Annie C. Y. Chang and Stanley N. Cohen, "Genome Construction Between Bacterial Species *In Vitro*: Replication and Expression of *Staphylococcus* Plasmid Genes in *Escherichia coli*," Proceedings of the National Academy of Sciences USA 71, no. 4 (1974): 1030–34; J. F. Morrow et al., "Replication of Transcription of Eukaryotic DNA in *Escherichia coli*," *Proceedings of the National Academy of Sciences USA* 71, no. 5 (1974): 1743–47.

15 *Mice were up next*: J. W. Gordon et al., "Genetic Transformation of Mouse Embryos by Microinjection of Purified DNA," *Proceedings of the National Academy of Sciences* 77, no. 12 (1980): 7380–84; J. W. Gordon and F. H. Ruddle, "Integration and Stable Germ Line Transmission of Genes Injected into Mouse Pronuclei," *Science* 214, no. 4526 (1981): 1244–46; F. Costantini and E. Lacy, "Introduction of a Rabbit β-globin Gene into the Mouse Germ Line," *Nature* 294, no. 5836 (1981): 92.

16 *green fluorescent protein*: Information about GFP, its discovery and history, is from Roger Y. Tsien, "The Green Fluorescent Protein," *Annual Review of Biochemistry* 67 (1998): 509–44.

16 *Zhiyuan Gong, a biologist*: Information on Gong's goals and work is from Alan Blake, in discussion with author, Austin, Texas, December 4, 2010; B. Ju et al., "Faithful Expression of Green Fluorescent Protein (GFP) in Transgenic Zebrafish Embryos under Control of Zebrafish Gene Promoters," *Developmental Genetics* 25, no. 2 (1999): 158–67; Zhiyuan Gong et al., "Development of Transgenic Fish for Ornamental and Bioreactor by Strong Expression of Fluorescent Proteins in the Skeletal Muscle," *Biochemical and Biophysical Research Communications* 308, no. 1 (August 15, 2003): 58–63.

17 *His team accomplished that feat*: Ju et al., "Faithful Expression of Green Fluorescent Protein (GFP)." The National University of Singapore wasn't the only university that had reinvented zebrafish; fluorescent versions of the creatures had also been created for various research projects at other labs around the world.

17 *In subsequent research*: Gong et al., "Development of Transgenic Fish for Ornamental and Bioreactor."

17 *Crockett vividly remembers*: Richard Crockett, e-mail message to author, January 31, 2012.

17 *In 1998, at the age of twenty-one . . . dot-com crash*: Alan Blake, e-mail message to author, April 2, 2012.

17 *As the two young men cast about*: Details about the early conversations that led to GloFish come from Alan Blake, discussion, December 2010, and in conversation with author via telephone, June 13, 2011.

17*n* *In 2005, Gong's team*: Z. Zeng et al., "Development of Estrogen-Responsive Transgenic Medaka for Environmental Monitoring of Endocrine Disrupters," *Environmental Science & Technology* 35 (2005): 9001–9008.

17*n* *In 2010, scientists at China's Fudan University*: H. Chen et al., "Generation of a Fluorescent Transgenic Zebrafish for Detection of Environmental Estrogens," *Aquatic Toxicology* 96 (2010): 53–61.

17*n* *Despite these advances*: Stephen Smith, "S. Korea Uses Goldfish to Test G20 Water; PETA Protests," CBS News, November 11, 2010, www.cbsnews.com/8301-503543_162-20022538-503543.html

18 *The pair founded Yorktown Technologies*: Details about the early days of Yorktown Technologies come from Blake, discussion, December 2010.

18 *"The ornamental fish industry"*: Blake, discussion, January 2010.

19 *When Herzog consulted*: Harold Herzog, "Forty-two Thousand and One Dalmatians: Fads, Social Contagion, and Dog Breed Popularity," *Society and Animals* 14, no. 4 (2006): 383–97; Hal Herzog, *Some We Love, Some We Hate, Some We Eat* (New York: HarperCollins, 2010), 117–21.

19 *In antiquity*: For more on our ancestors' interest in exotic animals, see Linda Kalof, *Looking at Animals in Human History* (London: Reaktion Books, 2007).

19 *Even the humble goldfish*: Information about the early history of the goldfish

comes from E. K. Balon, "About the Oldest Domesticates Among Fishes," *Journal of Fish Biology* 65, Supplement A (2004): 1–27.

19 *As goldfish grew in popularity*: Ibid.

20 *A 2007 study*: David L. Stokes, "Things We Like: Human Preferences Among Similar Organisms and Implications for Conservation," *Human Ecology* 35, no. 3 (2007): 361–69.

20 *We've bred canaries*: Companion Animal Welfare Council, *Breeding and Welfare in Companion Animals* (UK: May 2006).

20 *And before GloFish were even*: Blake, discussion, December 2010; Eric M. Hallerman, discussion, September 2011; Svein A. Fossa, "Man-Made Fish: Domesticated Fishes and Their Place in the Aquatic Trade and Hobby," *Ornamental Fish International Journal* 44 (February 2004): 1–16.

20*n* *Scientists have created beagles*: S. G. Hong et al., "Generation of Red Fluorescent Protein Transgenic Dogs," *Genesis* 47 (May 2009): 314–22.

21 *origin of the modern Labradoodle*: "Labradoodle History," International Labradoodle Association, accessed March 6, 2012, www.ilainc.com/Labradoodle History.html; "About the Labradoodle," International Labradoodle Association, accessed March 6, 2012, www.ilainc.com/AboutTheLabradoodle.html; Miriam Fields-Babineau, *Labradoodle: Comprehensive Owner's Guide* (Allenhurst, NJ: Kennel Club Books, 2006), 9–10; Margaret Bonham, *Labradoodles: A Complete Pet Owner's Manual* (New York: Barron's Educational Series, 2007), 8.

22 *"Through advances in genetic . . ."*: *Wildcard—Genetically Modified Pets* (Washington, DC: Social Technologies, 2007).

22 *A company called Felix Pets*: "Felix Pets," Felix Pets, LLC, accessed May 27, 2012, www.felixpets.com/welcome.html.

22 *"If we're going to come up . . ."*: Alan Beck, in discussion with author via telephone, November 11, 2009.

22 *The Food and Drug Administration considers*: In 2009, the FDA issued a document outlining how it planned to regulate transgenic organisms: Center for Veterinary Medicine, U.S. Food and Drug Administration, *Guidance for Industry: Regulation of Genetically Engineered Animals Containing Heritable Recombinant DNA Constructs* (Rockville, MD: January 15, 2009), available at www.fda.gov/downloads/AnimalVeterinary/GuidanceComplianceEnforcement/GuidanceforIndustry/UCM113903.pdf. Alison L. Van Eenennaam et al., *The Science and Regulation of Food from Genetically Engineered Animals* (Council for Agricultural Science and Technology, June 2011), available at www.cast-science.org/publications/?the_science_and_regulation_of_food_from_genetically_engineered_animals&show=product&productID=21628. Hallerman, discussion, February 2011.

22*n* *Another company, Lifestyle Pets*: Information about the company, and its claims, products, and pricing comes from "Lifestyle Pets," Lifestyle Pets, accessed June 17, 2011, www.allerca.com. See also Michael Hopkin, "Allergy-free Pets Surprisingly Simple," *Nature News*, September 26, 2006, www.nature.com/news/2006/060926/full/news060925-5.html.

22n *controversy has long swirled*: For more on the controversy, see Kerry Grens, "FelisEnigmaticus," *The Scientist*, January 1, 2007, http://classic.the-scientist.com/article/home/39383/.

23 *In 1975, they drew up*: Paul Berg et al., "Summary Statement of the Asilomar Conference on Recombinant DNA Molecules," *Proceedings of the National Academy of Sciences USA* 72, no. 6 (1975): 1981–84. See also Paul Berg and Maxine Singer, "The Recombinant DNA Controversy: Twenty Years Later," *Proceedings of the National Academy of Sciences USA* 92 (September 1995): 9011–13.

23 *The National Institutes of Health issued guidelines*: "About Recombinant DNA Advisory Committee (RAC)," National Institutes of Health, accessed March 28, 2012, http://oba.od.nih.gov/rdna_rac/rac_about.html; "NIH Guidelines for Research Involving Recombinant DNA Molecules," National Institutes of Health, accessed March 28, 2012, http://oba.od.nih.gov/oba/rac/Guidelines/NIH_Guidelines.htm.

23 *ecologists continue to worry*: There has been a lot written on the potential environmental risks of genetically engineered fish. My information comes from a number of sources, including Hallerman, discussion, September 2011; John A. Beardmore and Joanne S. Porter, *Genetically Modified Organisms and Aquaculture* (Rome: Food and Agriculture Organization of the United Nations, 2003), 3–4; *Future Fish: Issues in Science and Regulation of Transgenic Fish* (Washington, DC: Pew Initiative on Food and Biotechnology, January 2003); Erik Stokstad, "Engineered Fish: Friend or Foe of the Environment?" *Science* 297 (September 13, 2002): 1797–99; Alison L. Van Eenennaam and Paul G. Olin, "Careful Risk Assessment Needed to Evaluate Transgenic Fish," *California Agriculture* 60 (July–September 2006): 126–31; Alison L. Van Eenennaam and William M. Muir, "Transgenic Salmon: A Final Leap to the Grocery Shelf," *Nature Biotechnology* 29 (2011): 706–10.

23 *This very possibility has been*: Information about the salmon comes from many sources, including: "AquAdvantage Fish," AquaBounty Technologies, Inc., accessed March 23, 2012, www.aquabounty.com/products/aquadvantage-295.aspx; "Frequently Asked Questions," AquaBounty Technologies, Inc., accessed March 23, 2012, www.aquabounty.com/technology/faq-297.aspx; Aqua Bounty Technologies, Inc, *Environmental Assessment for AquAdvantage® Salmon* (submitted to the Center for Veterinary Medicine, US Food and Drug Administration, August 25, 2010), available at www.fda.gov/downloads/AdvisoryCommittees/ . . . /UCM224760.pdf; Veterinary Medicine Advisory Committee, Center for Veterinary Medicine, Food and Drug Administration, *Briefing Packet: AquAdvantage Salmon* (September 20, 2010), available at www.fda.gov/downloads/AdvisoryCommittees/ . . . /UCM224762.pdf; *Future Fish: Issues in Science and Regulation of Transgenic Fish*; and Hallerman, discussion, September 2011. The initial scientific work that led to the AquAdvantage salmon is Shao Jun Du et al., "Growth Enhancement in Transgenic Atlantic Salmon by the Use of an 'All Fish' Chimeric Growth Hormone Gene Construct," *Nature Biotechnology* 10 (1992): 176–81.

24 *AquaBounty is building several security*: Hallerman, discussion, September 2011; Alison Van Eenennaam, in discussion with author via telephone, February 8, 2012; Eenennaam and Muir, "Transgenic Salmon"; Van Eenennaam et al., *The Science and Regulation of Food from Genetically Engineered Animals.*

24 *Though many scientists*: Van Eenennaam et al., *The Science and Regulation of Food from Genetically Engineered Animals*; Hallerman, discussion, September 2011.

24–25 *The company first approached . . . the market*: Van Eenennaam, discussion; Van Eenennaam and Muir, "Transgenic Salmon."

25 *As Alan Blake prepared . . . to the environment*: Blake, discussion, December 2010.

25 *Wild zebrafish . . . successful at reproducing*: Hallerman, discussion, February 2011. Van Eenennaam and Olin, "Careful Risk Assessment Needed to Evaluate Transgenic Fish"; and Blake, "GloFish—The First Commercially Available Biotech Animal." Data and analyses of GloFish risks appear in a memorandum written by Sonke Mastrup, acting director of the California Department of Fish and Game, and in a series of letters written by experts to Alan Blake. They include Sonke Mastrup to Robert R. Treanor, memorandum, November 25, 2003, available at www.glofish.com/science/CA.Fish.Game.Recommendation.pdf; Eric M. Hallerman to Alan Blake, September 18, 2003, available at http://glofish.com/science/Hallerman%20Analysis%20of%20Fluorescent%20Zebra%20Fish.pdf; Jeffrey J. Essner to Alan Blake, October 14, 2003, available at http://glofish.com/science/Analysis%20of%20Fluorescent%20Zebra%20Fish%20Temperature%20Sensitivity.pdf; Perry B. Hackett to Alan Blake, August 18, 2003, available at http://glofish.com/science/Hackett%20Analysis%20of%20Fluorescent%20Tropical%20Fish.pdf; William Muir to Alan Blake, November 16, 2003, available at http://glofish.com/science/Muir%20Analysis%20of%20Fluorescent%20Zebra%20Fish.pdf; and Zhiyuan Gong to Alan Blake, September 3, 2003, available at http://glofish.com/science/Gong%20Analysis%20of%20Fluorescent%20Zebra%20Fish.pdf.

25 *"What are the odds . . . "*: Perry Hackett, in discussion with author via telephone, February 4, 2011.

25 *Federal officials didn't register*: Blake, discussion, December 2010.

26 *then California caught him by surprise*: Blake, discussion, December 2010.

26 *The state's Fish and Game Commission*: Van Eenennaam and Olin, "Careful Risk Assessment Needed to Evaluate Transgenic Fish."

26 *National Public Radio to Al-Jazeera*: Blake, discussion, December 2010.

26 WHEN FISH FLUORESCE: James Gorman, "When Fish Fluoresce, Can Teenagers Be Far Behind?" *New York Times*, December 2, 2003.

26 *"These are techniques that . . . "*: Richard Twine, in discussion with author via telephone, November 11, 2009.

27 *when it convened to discuss GloFish*: A video of the proceedings is available online at CAL-SPAN: California State Meetings Webcast Video. Videos of California Fish and Game Commission proceedings are available at www.cal-span.org/cgi-bin/media.pl?folder=CFG. The video of the December 3, 2003,

meeting can be downloaded directly at mms://media.cal-span.org/calspan/Video_Files/CFG/CFG_03-12-03/CFG_03-12-03.wmv. All details about what occurred at the meeting, and quotations cited, come directly from the video.

27 *According to public opinion polls*: "Recent Findings," Mellman Group, Inc., and Public Opinion Strategies, Inc., to the Pew Initiative on Food and Biotechnology, memorandum, November 7, 2003, available at www.pewtrusts.org/uploadedFiles/wwwpewtrustsorg/Public_Opinion/Food_and_Biotechnology/2005summary.pdf.

28 *"a fairly trivial use of technology" . . . "no harm being done"*: Hallerman, discussion, February 2011.

28 *Those ornamental goldfish varieties*: Fossa, "Man-Made Fish"; Companion Animal Welfare Council, *Breeding and Welfare.*

28*n* *The breed's massive head*: *A Healthier Future for Pedigree Dogs: The Report of the APGAW Inquiry into the Health and Welfare Issues Surrounding the Breeding of Pedigree Dogs* (London: Associate Parliamentary Group for Animal Welfare, November 2009).

28*n* *Their snouts are so short*: James A. Serpell, "Anthropomorphism and Anthropomorphic Selection—Beyond the 'Cute Response,'" *Society & Animals* 11, no. 1 (2003): 83–100.

28*n*–29*n* *These breathing difficulties*: Nicola Rooney and David Sargan, "Pedigree Dog Breeding in the UK: A Major Welfare Concern?" (UK: Royal Society for the Prevention of Cruelty to Animals, 2009).

29 *"Because tropical aquarium fish . . ."*: United States Food and Drug Administration, "FDA Statement Regarding Glofish," December 9, 2003, available at www.fda.gov/AnimalVeterinary/DevelopmentApprovalProcess/GeneticEngineering/GeneticallyEngineeredAnimals/ucm161437.htm

29 *GloFish hit pet stores in January 2004*: Blake, discussion, December 2010.

29 *filed a lawsuit*: Complaint, *International Center for Technology Assessment v. Thompson*, No. 1:04-CV-000 62 (D.D.C. January 14, 2004), available at www.centerforfoodsafety.org/pubs/GloFishComplaint1.14.2004.pdf.

29*n* *"If bulldogs were the products . . ."*: Serpell, "Anthropomorphism and Anthropomorphic Selection—Beyond the 'Cute Response.'"

30 *"aesthetic injury from viewing . . . "*: Ibid.

30 *are a hit . . . major pet store chains*: Blake, discussions, January and December 2010. Blake declined to provide specific sales figures but said, "Our fish are among the most popular in the marketplace."

30 *the Taiwanese company Taikong*: The company's site can be found at www.azoo.com.tw/

30 *Though he'd love to sell*: Alan Blake, in discussion with author via telephone, September 6, 2011, discussion, December 2010.

30 *At first, Yorktown Technologies*: Blake, discussion, December 2010, and e-mail to author, February 1, 2012.

30*n* *Yorktown Technologies conducted . . . sale of the fish, Blake says*: Alan Blake, e-mail to author, August 2, 2012. Alan Blake, e-mail to author, August 20, 2012.

31 *"We have e-mails . . ."*: Blake, discussion, January 2010.

31 *In one survey, 40 percent*: "Recent Findings," Mellman Group, Inc., and Public Opinion Strategies, Inc., to the Pew Initiative on Food and Biotechnology, memorandum, November 7, 2003, available at www.pewtrusts.org/uploaded Files/wwwpewtrustsorg/Public_Opinion/Food_and_Biotechnology/2005 summary.pdf.

31 *"Biotechnology is often demonized"*: Blake, discussion, January 2010.

32 *"You'd think they were . . ."*: Alan Blake, in discussion with author via telephone, October 13, 2010.

2. Got Milk?

33 *Instead, most researchers*: James Murray, in discussion with author via telephone, September 1, 2011; Murray and Elizabeth Maga, in discussion with author, Davis, California, January 24, 2012.

33 *The trouble is that it's difficult*: Louis-Marie Houdebine, "Production of Pharmaceutical Proteins by Transgenic Animals," *Comparative Immunology, Microbiology and Infectious Diseases* 32 (2009): 107–21; Michael K. Dyck, "Making Recombinant Proteins in Animals—Different Systems, Different Applications," *TRENDS in Biotechnology* 21, no. 9 (2003): 394–99; Murray, discussion, September 2011; Murray and Maga, discussion, January 2012.

33 *So the creation*: Murray, discussion, September 2011; Murray and Maga, discussion, January 2012.

34 *Throughout the 1980s . . . in their milk*: C. W. Pittius et al., "A Milk Protein Gene Promoter Directs the Expression of Human Tissue Plasminogen Activator cDNA to the Mammary Gland in Transgenic Mice," *Proceedings of the National Academy of Sciences* 85 (1988): 5874–78; K. Gordon et al., "Production of Human Tissue Plasminogen Activator in Transgenic Mouse Milk," *Bio/Technology* 5 (1987): 1183–87; G. Wright et al., "High Level Expression of Active Human Alpha-1-Antitrypsin in the Milk of Transgenic Sheep," *Nature Biotechnology* 9 (1991): 830–34.

34 *That all changed with ATryn . . . genetically engineered goats*: Information on antithrombin and ATryn is from many sources, including "Hereditary Antithrombin Deficiency," U.S. National Library of Medicine, National Institutes of Health, accessed April 17, 2012, http://ghr.nlm.nih.gov/condition/heredi tary-antithrombin-deficiency; Jim Kling, "First US Approval for a Transgenic Animal Drug," *Nature Biotechnology* 27, no. 4 (2009): 302–304; "Summary Basis for Regulatory Action—ATryn," U.S. Food and Drug Administration, accessed February 4, 2009, www.fda.gov/BiologicsBloodVaccines/BloodBlood Products/ApprovedProducts/LicensedProductsBLAs/FractionatedPlasma Products/ucm134048.htm; "FDA Approves Orphan Drug ATryn to Treat Rare Clotting Disorder," U.S. Food and Drug Administration, February 6, 2009, www.fda.gov/NewsEvents/Newsroom/PressAnnouncements/2009/ucm 109074.htm; "ATryn® (Antithrombin [Recombinant]) Approved by the FDA,"

GTC Biotherapeutics, February 6, 2009, www.gtc-bio.com/pressreleases/pr020609.html. You can download many of the regulatory documents, and much more information, at "ATryn," U.S. Food and Drug Administration, accessed April 17, 2012, www.fda.gov/BiologicsBloodVaccines/BloodBloodProducts/ApprovedProducts/LicensedProductsBLAs/FractionatedPlasmaProducts/ucm134042.htm.

34 *To create its special herd*: The procedure GTC used to make its transgenic goats is recounted in many places, including Kling, "First US Approval for a Transgenic Animal Drug"; "How It Works," GTC Therapeutics, accessed February 10, 2012, www.gtc-bio.com/science.html; and "Questions by Scientists," GTC Therapeutics, accessed February 10, 2012, www.gtc-bio.com/science/questions.html.

35 *ATryn hit the market*: Kling, "First US Approval for a Transgenic Animal Drug."

35 *"milking parlors" on GTC's 300-acre farm*: "The GTC Biotherapeutics Production Facility," GTC Biotherapeutics, accessed February 10, 2012, www.gtc-bio.com/science/production.html.

35 *more than a kilogram*: "Questions by Scientists," GTC Therapeutics.

35 *Ruconest, a drug produced*: For more, see "Ruconest," Pharming Group NV, accessed April 5, 2012, www.pharming.com/index.php?act=prod.

35*n* *The EU approved*: Ibid.

35*n* *Scientists have also created . . . on a protein*: C. L. Keefer, "Production of Bioproducts Through the Use of Transgenic Animal Models," *Animal Reproduction Science* 82–83 (2004): 5–12; Houdebine, "Production of Pharmaceutical Proteins by Transgenic Animals"; Dyck, "Making Recombinant Proteins in Animals"; James Murray, in discussion with author via telephone, February 29, 2012; and Murray, discussion, 2011.

36 *The ailment's global toll*: James D. Murray et al., "Current Status of Transgenic Animal Research for Human Health Applications," in "24th Brazilian Embryo Technology Society (SBTE) Annual Meeting," supplement 2, *Acta Scientiae Veterinariae* 38 (2010): s627–32.

36 *The evidence now suggests*: Dottie R. Brundige et al., "Lysozyme Transgenic Goats' Milk Influences Gastrointestinal Morphology in Young Pigs," *Journal of Nutrition* 138 (2008): 921–26; Lene Schack-Nielson and Kim F. Michaelsen, "Advances in Our Understanding of the Biology of Human Milk and Its Effects on the Offspring," *The Journal of Nutrition* 137, no. 2 (2007): 503S–510S.

36 *Some of these effects can last*: P. W. Howie et al., "Protective Effect of Breast Feeding Against Infection," *BMJ* 300 (January 6, 1990): 11–16.

36 *One of the compounds*: Murray, discussion, 2011.

36–37 *Lysozyme is naturally present . . . other animals*: Elizabeth A. Maga et al., "Consumption of Milk from Transgenic Goats Expressing the Human Lysozyme in the Mammary Gland Results in the Modulation of Intestinal Microflora," *Transgenic Research* 15 (2006): 515–19.

37 *Infant formula, which is usually*: Elizabeth Maga, in discussion with author via telephone, April 16, 2012; Clifford W. Lo and Ronald E. Kleinman, "Infant Formula, Past and Future: Opportunities for Improvement," *American Journal of Clinical Nutrition* 63 (1996): 646S–650S.

37 *Like the scientists at GTC . . . mulberry leaves*: E. A. Maga et al., "Production and Processing of Milk from Transgenic Goats Expressing Human Lysozome in the Mammary Gland," *Journal of Dairy Science* 89 (2006): 518–24. Murray and Maga, discussion, January 2012.

37 *The barn is home to 150 assorted goats*: Murray, discussion, February 2012.

37–38 *the "founder female"*: Murray and Maga, discussion, January 2012.

38 *They'll make up . . . after they give birth*: Murray, discussion, January 2012.

38 *1,000 percent more*: Maga et al., "Production and Processing of Milk from Transgenic Goats."

38 *They've also demonstrated . . . diarrheal disease*: Ibid.

38 *They also had stronger immune*: Brundige et al., "Consumption of Pasteurized Human Lysozyme"; C. A. Cooper et al., "Lysozyme Transgenic Goats' Milk Positively Impacts Intestinal Cytokine Expression and Morphology," *Transgenic Research* 20, no. 6 (2011): 1235–43.

38 *And when the researchers* tried: Brundige et al., "Lysozyme Transgenic Goats' Milk."

39 *in September 2011, they asked*: Murray and Maga, discussion, January 2012.

39 *The compound is well studied*: Murray, discussion, January 2012.

39 *"You've been eating lysozyme . . . "*: Ibid.

39 *Still, Murray and Maga aren't sure*: Murray and Maga, discussion, 2011.

39 *Murray and Maga are hedging their bets*: Details about the Brazil collaboration are from Murray and Maga, discussions, 2011 and January 2012; "U.S.-Brazilian Research Team to Tackle Deadly Intestinal Diseases with Genetically Enhanced Goats' Milk," UC Davis, accessed April 5, 2012, http://caes.ucdavis.edu/NewsEvents/web-news/2009/march-2009/u.s.-brazilian-research-team-to-tackle-deadly-intestinal-diseases-with-genetically-enhanced-goats2019-milk.

39 *which is among a handful*: Murray and Maga, discussion, January 2012; Van Eenennaam, discussion; Nicolas Rigaud, *Biotechnology: Ethical and Social Debates* (OECD International Futures Programme, February 2008); "Brazil's Biotech Boom," *Nature* 466, no. 7304 (2010): 295.

40 *The region is home . . . fifth birthday*: "U.S.-Brazilian Research Team," UC Davis.

40 *Assuming the human trials . . . levels of lysozyme*: Murray and Maga, discussion, January 2012; Maga, discussion, April 2012.

40*n* *Murray and Maga haven't quite decided*: Elizabeth Murray, e-mail message to author, June 13, 2012.

41 *In 1975, just as . . . upon other species*: Peter Singer, *Animal Liberation*, rev. ed. (1975; New York: HarperPerennial, 2009).

42 *"Animals that have previously been . . ."*: Richard Twine, in discussion with author via telephone, February 21, 2012.

42 *in the United States alone, ten people*: "Xenotransplantation," U.S. Food and Drug Administration, accessed March 9, 2012, www.fda.gov/BiologicsBloodVaccines/Xenotransplantation/default.htm.

42–43 *Throughout the twentieth century . . . than two months*: *Animal-to-Human Transplants: The Ethics of Xenotransplantation* (London: Nuffield Council on Bioethics, 1996); L. L. Bailey et al., "Baboon-to-Human Cardiac Xenotransplantation in a Neonate," *Journal of the American Medical Association* 254 (1985):3321–29.

42*n* *Since 1995, the number . . . involved GM animals*: Home Office, *Statistics of Scientific Procedures on Living Animals, Great Britain 2010* (London: The Stationery Office Limited, 2011), available at www.homeoffice.gov.uk/publications/science-research-statistics/research-statistics/other-science-research/spanimals10/spanimals10?view=Binary.

42*n* *Japan's labs are home*: Kenichi Yagami et al., "Survey of Live Laboratory Animals Reared in Japan (2009)," *Experimental Animals* 59, no. 4 (2010): 531–35.

42*n* *the USDA issues an annual report*: U.S. Department of Agriculture, *Annual Report Animal Usage by Fiscal Year, 2010* (July 27, 2011), available for download at www.aphis.usda.gov/animal_welfare/efoia/downloads/2010_Animals_Used_In_Research.pdf.

42*n* *Take the popular procedure*: Nicole L. Miller and Brant R. Fulmer, "Injection, Ligation and Transplantation: The Search for the Glandular Fountain of Youth," *The Journal of Urology* 177 (June 2007): 2000–2005.

43 *Surgeons can keep pig valves*: Nuffield Council on Bioethics, *Animal-to-Human Transplants*, 26.

43 *Enter genetic engineering*: Information on how genetic engineering could make xenotransplantation a reality is from ibid.; L. Paterson et al., "Application of Reproductive Biotechnology in Animals: Implications and Potentials. Applications of Reproductive Cloning," *Animal Reproductive Science* 79, nos. 3–4 (2003): 137–43; Murray et al., "Current Status of Transgenic Animal Research"; Desmond S. T. Nicholl, *An Introduction to Genetic Engineering*, 3rd ed. (Cambridge, UK: Cambridge University Press, 2008), 276; "Xenotransplantation," U.S. FDA.

44 *Sixty-two percent of Americans*: Heather Mason Kiefer, "Americans Unruffled by Animal Testing," Gallup, Inc., May 25, 2004, www.gallup.com/poll/11767/americans-unruffled-animal-testing.aspx.

44 *Monstrous mash-ups*: For overviews of the public reaction to GM animals, see Phil Macnaghten, "Animals in Their Nature: A Case Study on Public Attitudes to Animals, Genetic Modification, and 'Nature,'" *Sociology* 38, no. 3 (2004): 533–51; E. F. Einsiedel, "Public Perceptions of Transgenic Animals," *Revue Scientifique et Technique* 24, no. 1 (2005): 149–57. For scholars' opinions on these same issues, see the collection of articles in *The American Journal of Bioethics* 3, no. 3 (2003). See also Bernard Rollin, *The Frankenstein Syndrome* (Cambridge, UK: Cambridge University Press, 1995).

44*n* *Polls reveal that between 97*: Herzog, *Some We Love*, 191.

44*n* *our appetite for meat*: Ibid.

45 *In a series of recent experiments*: Esmail D. Zanjani et al., "Generation of Functional Humanized Liver in Sheep by Bone Marrow Cells," *The Journal of Federation of American Societies for Experimental Biology* 23 (April 2009): 186.3; Judith A. Airey et al., "Human Mesenchymal Stem Cells Form Purkinje Fibers in Fetal Sheep Heart," *Circulation* 109 (2004): 1401–1407; Adel Ersek et al., "Persistent Circulating Human Insulin in Sheep Transplanted In Utero with Human Mesenchymal Stem Cells," *Experimental Hematology* 38, no. 4 (2010): 311–20.

45*n* *In 1927, Ivanov . . . this plan*: Kirill Rossiianov, "Beyond Species: Il'ya Ivanov and His Experiments on Cross-Breeding," *Science in Context* 15, no. 2 (2002), 277–316.

46 *Louisiana and Arizona*: Louisiana Rev. Stat. 14:89.6 (2009). Arizona Rev. Stat. 36–2311 (2010).

46 *Human-Animal Hybrid Prohibition Act*: Human-Animal Hybrid Prohibition Act of 2009, S. 1435, 111th Cong. (2009). Brownback introduced the bill in 2009, but the Senate never took it up for discussion.

46 *It's interesting to note that we rarely hear*: In fact, the flip side of the argument occurred to me only after I read it in the Academy of Medical Sciences report *Animals Containing Human Material*.

46 *Animal cognition has much in common*: Academy of Medical Sciences, *Animals Containing Human Material*, 46–48.

46 *In 2009, German researchers*: Wolfgang Enard et al., "A Humanized Version of Foxp2 Affects Cortico-basal Ganglia Circuits in Mice," *Cell* 135, no. 5 (2009): 961–71.

47 *These sticky philosophical questions*: Academy of Medical Sciences, *Animals Containing Human Material*, 9.

47 *Similarly, in the United States*: Ibid., 103–104.

47 *biologists argue over just what*: Jason Scott Robert and Francoise Baylis, "Crossing Species Boundaries," *American Journal of Bioethics* 3, no. 3 (Summer 2003): 1–13.

47*n* *African scientists and policymakers . . . a form of cannibalism*: Murray, discussion, February 2012.

47*n* *"the great majority . . . no novel issues"*: Academy of Medical Sciences, *Animals Containing Human Material*, 7.

48 *Different species of bacteria . . . other animals*: J. C. Dunning Hotopp, "Horizontal Gene Transfer Between Bacteria and Animals," *Trends in Genetics* 27, no. 4 (2011): 157–63.

48 *The parasite that causes Chagas' disease*: Mariana M. Hecht et al., "Inheritance of DNA Transferred from American Trypanosomes to Human Hosts," *PLoS ONE* 5, no. 2 (2010): e9181.

48 *pea aphids have borrowed*: Nancy A. Moran and Tyler Jarvik, "Lateral Transfer of Genes from Fungi Underlies Carotenoid Production in Aphids," *Science* 328, no. 5978 (2010): 624–27.

48 *and it's what makes us recoil*: For more on how the yuck factor manifests itself in both cases—toilet water and biotechnology—see Charles W. Schmidt, "The Yuck Factor: When Disgust Meets Discovery," *Environmental Health Perspectives* 116, no. 12 (2008): A524–27.

49 *"In critical cases . . . our humanity"*: Leon R. Kass, "The Wisdom of Repugnance," *The New Republic*, June 2, 1997, 20. For another take, see Mary Midgley, "Biotechnology and Monstrosity: Why We Should Pay Attention to the 'Yuk Factor,'" *Hastings Center Report* 30, no. 5 (2000): 7–15.

49 *"appears to be irrational . . ."*: Academy of Medical Sciences, *Animals Containing Human Material*, 70.

49 *seeing an interracial couple*: I naïvely thought that I had come up with this great parallel, but it turns out that the disgust once common at the sight of interracial couples has been cited by many ethicists as evidence that there is not always wisdom in repugnance.

49 *"conservation of welfare"*: Rollin, *The Frankenstein Syndrome*; Rollin, in discussion with author via telephone, February 13, 2012.

49 *"If you're going to modify . . ."*: Rollin, discussion.

50 *the "Beltsville pig"*: Information on the Beltsville pig is from Rollin, *The Frankenstein Syndrome*. Committee on Defining Science-Based Concerns Associated with Products of Animal Biotechnology et al., *Animal Biotechnology: Science Based Concerns* (Washington, DC: The National Academies Press, 2002), 98.

50 *Our ability to limit . . . unusual health problems*: Murray and Maga, discussion, January 2012; Murray, discussion, February 2012.

50 *the FDA examined seven generations*: "FDA Approves Orphan Drug ATryn," U.S. FDA.

51 *As Rollin put it in his book*: Rollin, *The Frankenstein Syndrome*.

51 *In fact, Murray and Maga's goats*: Murray, discussion, 2011; Murray and Maga, discussion, January 2012; Maga et al., "Production and Processing of Milk from Transgenic Goats."

51 *Several labs, for instance, have created*: Jürgen A. Richt et al., "Production of Cattle Lacking Prion Protein," *Nature Biotechnology* 25 (2007): 132–38; Michael C. Golding et al., "Suppression of Prion Protein in Livestock by RNA Interference," *PNAS* 103, no. 14 (2006): 5285–90.

52 *Several teams of Chinese scientists*: Richard Gray, "Cows Genetically Modified to Produce Healthier Milk," *The Telegraph*, June 17, 2012, www.telegraph.co.uk/science/science-news/9335762/Cows-genetically-modified-to-produce-healthier-milk.html.

52 *Researchers in some labs*: Houdebine, "Production of Pharmaceutical Proteins by Transgenic Animals"; Dyck, "Making Recombinant Proteins in Animals."

52 *a whopping 9 grams of protein*: Dyck, "Making Recombinant Proteins in Animals."

52 *A team of Japanese biologists*: Masahiro Tomita et al., "Transgenic Silkworms Produce Recombinant Human Type III Procollagen in Cocoons," *Nature Biotechnology* 21, no. 1 (2003): 52–56.

52 *one hen can lay*: A. J. Harvey and R. Ivarie, "Validating the Hen as a Bioreactor for the Production of Exogenous Proteins in Egg White," *Poultry Science* 82, no. 6 (2003): 927–30.

52 *Scientists at Scotland's Roslin Institute*: S. G. Lillico et al., "Oviduct-Specific Expression of Two Therapeutic Proteins in Transgenic Hens," *Proceedings of the National Academy of Sciences* 104, no. 6 (2007): 1771–76; Helen Sang, in discussion with author via telephone, August 31, 2011.

52 *Before long, we could*: For more on the egg as a vehicle for therapeutic proteins, see James N. Petitte and Paul E. Mozdziak, "The Incredible, Edible, and Therapeutic egg," *Proceedings of the National Academy of Sciences* 104, no. 6 (2007): 1739–40.

52 *"The way we've been making transgenics . . ."*: Van Eenennaam, discussion.

52 *Today, scientists are far better*: Ibid.; Murray and Maga, discussion, January 2012.

53 *in 2010, the biologist J. Craig Venter*: Daniel G. Gibson, et al., "Creation of a Bacterial Cell Controlled by a Chemically Synthesized Genome," *Science* 329 (July 2, 2010): 52–56. Elizabeth Pennisi, "Synthetic Genome Brings New Life to Bacterium," *Science* 328 (May 21, 2010): 958–59.

53 *In 2012, Canadian researchers*: Andrew Pollack, "Move to Market Gene-Altered Pigs in Canada Is Halted," *New York Times*, April 3, 2012, www.nytimes.com/2012/04/04/science/gene-altered-pig-project-in-canada-is-halted.html?hpw.

53 *Researchers at the University*: S. P. Golovan et al., "Pigs Expressing Salivary Phytase Produce Low-Phosphorus Manure," *Nature Biotechnology* 19, no. 8 (2001): 741–45. For a less-technical overview, see "Enviropig," University of Guelph, accessed March 9, 2012, www.uoguelph.ca/enviropig/.

54 *the scientific team was unable*: Pollack, "Move to Market Gene-Altered Pigs."

54 *the animals were euthanized . . . laboratory environment*: Sarah Schmidt, "Genetically Engineered Pigs Killed after Funding Ends," Postmedia News, June 22, 2012, www.canada.com/technology/Genetically+engineered+pigs+killed+after+funding+ends/6819844/story.html.

54 *"I don't think anybody . . ."*: Murray, discussion, January 2012.

3. Double Trouble

57 *Tabouli and Baba Ganoush*: "At Play with Firm's Clone Kittens," BBC News, August 9, 2004, http://news.bbc.co.uk/2/hi/science/nature/3548210.stm.

57 *Lancelot Encore*: Michael Inbar, "Encore! Couple Spend $155,000 to Clone Dead Dog," MSNBC.com, January 28, 2009.

58 *The world of cloning changed*: Information about Dolly comes from Ian Wilmut et al., "Viable Offspring Derived from Fetal and Adult Mammalian Cells," *Nature* 385 (February 27, 1997): 810–13; Ian Wilmut and Roger Highfield, *After Dolly: The Uses and Misuses of Human Cloning* (New York: W. W. Norton & Co., 2006).

58 *When Dolly was born*: Nicholl, *An Introduction to Genetic Engineering.*

58 *Like just about everything . . . first in the nation*: National Agricultural Statistics Service, United States Department of Agriculture, *Census of Agriculture State Profile*: *Texas* (USDA, 2007), available at www.agcensus.usda.gov/Publications/2007/Online_Highlights/County_Profiles/Texas/cp99048.pdf.

58 *The school has more than 700 acres*: The O. D. Butler, Jr., Animal Science Complex, which includes sheep, goat, and beef centers, consists of 580 acres; a separate horse center is 120 acres. "O. D. Butler, Jr., Animal Science Complex," Texas A&M University, accessed April 1, 2012, http://animalscience.tamu.edu/about/facilities/butler-ansc-complex/index.htm; "Horse Center," Texas A&M University, accessed April 1, 2012, http://animalscience.tamu.edu/about/facilities/horse-center/index.htm.

59 *In the years that followed*: Duane Kraemer, in discussion with author, College Station, Texas, December 5, 2010.

59 *Six months after . . . silky gray coat*: Ibid.; Mark Westhusin, in discussion with author via telephone, May 26, 2011; "Missy: Our Inspiration," Genetic Savings & Clone, April 27, 2006, www.savingsandclone.com/about_us/missy.html (site discontinued, accessed via Internet Archive on March 26, 2012, http://web.archive.org/web/20060427111502/www.savingsandclone.com/about_us/missy.html); "Background: Missyplicity Project," BioArts International, January 31, 2009, http://bestfriendsagain.com/missyplicity/index.html (site discontinued, accessed via Internet Archive on March 26, 2012, http://web.archive.org/web/20090131060948/http://bestfriendsagain.com/missyplicity/index.html); "About the Original," BioArts International, February 3, 2009, http://bestfriendsagain.com/missyplicity/original.html (site discontinued, accessed via Internet Archive on March 26, 2012, http://web.archive.org/web/20090203144140/http://bestfriendsagain.com/missyplicity/original.html); John Woestendiek, *Dog, Inc.* (New York: Penguin, 2010), 96.

59 *Missy belonged to Joan Hawthorne*: Woestendiek, *Dog, Inc.*

59 *After considering a number of labs . . . cloning effort*: Westhusin, discussion, May 2011, and e-mail message to author, February 22, 2012; Kraemer, discussion, December 2010.

59 *$3.7 million*: Lou Hawthorne, "A Project to Clone Companion Animals," *Journal of Applied Animal Welfare Science* 5, no. 3 (2002): 229–31.

59 *When the Missyplicity Project*: Kraemer, discussion, December 2010.

59 *"Millions of people . . ."*: Hawthorne, "A Project to Clone Companion Animals."

59 *Many of us consider*: Michael Schaffer, *One Nation Under Dog* (New York: Henry Holt and Co., 2009), 18. The book also has much more about the rising status of pets in our society and all the attendant consequences.

60 *off to the afterlife in style*: Schaffer, *One Nation Under Dog*, also has a great look at the burgeoning pet death industry.

60 *On February 16, 2000*: Hawthorne, "A Project to Clone Companion Animals."

60 *"a futuristic stocking stuffer . . ."*: "Genetic Savings & Clone Gift Certificates: The Perfect 21st Century Stocking Stuffer," Genetic Savings & Clone, Novem-

ber 20, 2001, www.savingsandclone.com/news/press_releases_05.html (site discontinued, accessed via Internet Archive on June 22, 2011, http://web.archive.org/web/20060510144138/www.savingsandclone.com/news/press_releases_05.html).

60 *The A&M team knew*: Mark Westhusin, e-mail message to author, June 12, 2012.

60 *so that's what they harvested*: Taeyoung Shin et al., "A Cat Cloned by Nuclear Transplantation," *Nature* 415 (February 21, 2002): 859.

61 *somatic-cell nuclear transfer*: The basics of somatic-cell nuclear transfer are covered in Wilmut and Highfield, *After Dolly*, 116–18; and Nicholl, *An Introduction to Genetic Engineering.*

61 *Westhusin and his team*: The procedure for cloning Rainbow is outlined in Shin et al., "A Cat Cloned by Nuclear Transplantation." (More-detailed procedures are available in the supplementary information that accompanies the paper online at www.nature.com/nature/journal/v415/n6874/suppinfo/nature723.html.)

61 *The egg, thus "tricked"*: This "tricked" description comes from Ian Wilmut, the scientist who led the Dolly team, in his description of somatic-cell nuclear transfer in Wilmut and Highfield, *After Dolly*, 118.

61 *The researchers ended up . . . was indeed Rainbow's clone*: Shin et al., "A Cat Cloned by Nuclear Transplantation."

61 *short for "Carbon Copy"*: Mark Westhusin, e-mail message, February 22, 2012.

61 *technically, clones produced*: Information about mitochondrial DNA and why animals created through SCNT are not quite identical to their donors comes from Wilmut and Highfield, *After Dolly*, 133–34; "Mitochondrial DNA," National Library of Medicine, National Institutes of Health, accessed March 12, 2012, http://ghr.nlm.nih.gov/chromosome/MT. Mark Westhusin, e-mail message to author, February 2, 2012.

62 *In creating Dolly, for instance . . . would be Dolly*: Wilmut et al., "Viable Offspring"; Wilmut and Highfield, *After Dolly*, 124.

62 *Dolly died at age six . . . took Dolly's life*: Wilmut and Highfield, *After Dolly*, 25–31.

62–63 *and assisted reproductive technologies . . . by other means*: Center for Veterinary Medicine, U.S. Food and Drug Administration, *Animal Cloning: A Risk Assessment* (Rockville, MD: January 8, 2008), available at www.fda.gov/AnimalVeterinary/SafetyHealth/AnimalCloning/UCM055489.

63 *That's what the FDA concluded . . . offspring appeared to be normal*: Ibid.

63 *"it is not possible . . ."*: Ibid., 10.

64 *"An egg knows how . . ."*: Westhusin, discussion, May 2011.

64 *Incomplete or flawed reprogramming*: For much more on reprogramming failures and abnormal gene expression in clones, see Center for Veterinary Medicine, *Animal Cloning: A Risk Assessment*, 59–92.

64 *Fortunately, CC was*: Shin et al., "A Cat Cloned by Nuclear Transplantation."

64 *For about a year, CC . . . claimed CC*: Kraemer, discussion, December 2010.

65 *He grew up on a dairy farm*: The details of Kraemer's background are from ibid.

65 *"CC has her own house . . ."*: Ibid.

66 *"We figured we should probably breed her . . ."*: Ibid.

66 *The matchmakers introduced CC . . . perfectly healthy*: Ibid.

66–67 *"I never thought . . . we had the lion"*: Shirley Kraemer, in discussion with author, College Station, Texas, December 5, 2010.

67 *So far, CC shows no signs*: Kraemer, discussion, December 2010.

67 *The mostly likely explanation . . . was turned off*: Ibid.; Kraemer, discussion, June, 2011.

67*n* *"a zoo brought a pair of lion cubs . . ."*: Kraemer, discussion, June 2011.

68 *they duplicated a Brahman bull named Chance*: Details about the cloning of Chance are from Westhusin, discussion, May 2011. The crew for the television version of *This American Life*, which was filming a segment on Fisher and his bulls, happened to be at the ranch the second time Second Chance attacked. Ira Glass, the show's host, interviewed Fisher in his hospital bed the next day. ("Reality Check," *This American Life*, season 1, episode 1, aired March 27, 2007).

68 *"Cloning is reproduction . . ."*: Westhusin, discussion, May 2011.

68 *For his part, Kraemer was thrilled*: Kraemer, discussion, December 2010.

68 *"People can be taken advantage of . . ."*: Kraemer, discussion, October 2009.

68 *"Nine Lives Extravaganza"*: "Nine Lives Extravaganza," Genetic Savings & Clone, August 8, 2004, http://savingsandclone.com/services/9lives.html (site discontinued, accessed via Internet Archive on March 26, 2012, http://web.archive.org/web/20040808043806/http://savingsandclone.com/services/9lives.html); John Suval, "Cloning for Cash: A&M's Pet Project Spawns a Company to Mix DNA with Possible IPOs," *Houston Press*, April 20, 2000; Wade Roush, "Genetic Savings & Clone: No Pet Project," *Technology Review*, March 2005; Ivan Oransky, "Cloning for Profit: Cloned Kittens Are Cute, but How Profitable Are Animal Cloning Companies?" *The Scientist*, January 31, 2005; Maryann Mott, "Cat Cloning Offered to Pet Owners," *National Geographic News*, March 25, 2004; "Bereaved Cat Owner Gets $50,000 Clone," *New York Times*, December 23, 2004, http://query.nytimes.com/gst/fullpage.html?res=9503E1DE1130F930A15751C1A9629C8B63.

68 *"If you feel that your kitten . . ."*: "Cat Cloning," Genetic Savings & Clone, April 27, 2006, http://savingsandclone.com/services/cat_cloning.html (site discontinued, accessed via Internet Archive, http://web.archive.org/web/20060427120819/http://savingsandclone.com/services/cat_cloning.html).

68 *a cloned Maine coon cat named Little Nicky*: "First-Ever Presentation of Pet Clone to Paying Client," Genetic Savings & Clone, December 23, 2004, www.savingsandclone.com/news/press_releases_11.html (site discontinued, accessed via Internet Archive, February 7, 2012, http://web.archive.org/web/20060510143749/www.savingsandclone.com/news/press_releases_11.html).

68 *"He is identical"*: Ibid.

69 *Despite the success*: Kraemer, discussion, December 2010; Woestendiek, *Dog, Inc.*
69 *The vagaries of the canine reproductive system . . . eggs out of her body*: Westhusin, discussion, May 2011, and e-mail message, June 2012.
69 *"The logistics of it . . ."*: Westhusin, discussion, May 2011.
69 *managed to coax two canines . . . stillborn pup*: Kraemer, discussion, December 2010.
69 *GSC shut down*: "Background: Missyplicity Project," BioArts International; Paul Elias, "Cat-Cloning Company to Close Its Doors," Associated Press, October 12, 2006.
69 *at the helm of a startup called BioArts International*: "BioArts Team: Lou Hawthorne," BioArts International, accessed June 24, 2011, http://bioartsinter national.com/team.htm.
69 *he connected with Hwang Woo Suk*: "Missy: Accomplished!" BioArts International, February 3, 2009, http://bestfriendsagain.com/missyplicity/missy.html (site discontinued, accessed via Internet Archive on March 26, 2012, http:// web.archive.org/web/20090203144135/http://bestfriendsagain.com/missy plicity/missy.html).
69 *the world's first cloned dog*: B. C. Lee, "Dogs Cloned from Adult Somatic Cells," *Nature* 436 (August 4, 2005): 641.
69–70 *Mira, Chingu, and Sarang*: "Missy: Accomplished!" BioArts International.
69*n* *Hwang has also been accused*: Information about Hwang's fall from grace is in David Cyranoski, "Verdict: Hwang's Human Stem Cells Were All Fakes," *Nature* 439 (January 12, 2006): 122–23; David Cyranoski, "Woo Suk Hwang Convicted, but Not of Fraud," *Nature* 461 (October 26, 2009); and "Timeline of a Controversy," *Nature*, December 19, 2005, www.nature.com/news/2005 /051219/full/news051219-3.html.
69*n* *"As a cloning company . . ."*: Ed Pilkington, "Dog Hailed as Hero Cloned by California Company," *Guardian*, June 18, 2009, www.guardian.co.uk/world /2009/jun/18/trakr-dog-september-11-clone.
70 *"Missy: Accomplished!"*: "Missy: Accomplished!" BioArts International.
70 *and noted that like*: "A Rose Is a Rose," BioArts International, February 3, 2009, www.bestfriendsagain.com/missyplicity/rose.html, (site discontinued, accessed via Internet Archive on March 26, 2012, http://web.archive.org/web /20090203042926/http://bestfriendsagain.com/missyplicity/rose.html).
70 *BioArts announced the "Best Friends Again" . . . a free copy*: "Auction Information," BioArts International, May 25, 2008, www.bestfriendsagain.com /auction/index.html (site discontinued, accessed via Internet Archive on June 24, 2011, http://web.archive.org/web/20080525201755/www.bestfriends again.com/auction/index.html); "Golden Clone Giveaway: Rules," BioArts International, June 4, 2008, www.bestfriendsagain.com/goldenclonegiveaway /rules.html (site discontinued, accessed via Internet Archive on June 24, 2011, http://web.archive.org/web/20080604090315/www.bestfriendsagain .com/goldenclonegiveaway/rules.html).

70 *50 million to 100 million animals*: Nuffield Council on Bioethics, *The Ethics of Research Involving Animals*, 7.

70*n* *the secret to the South Koreans' success . . . sold for their meat*: Westhusin, discussion, May 2011; Mark Walton, in discussion with author, Austin, Texas, December 3, 2010; Lou Hawthorne, "Six Reasons We're No Longer Cloning Dogs," BioArts International, September 10, 2009, accessed June 2, 2011, www.bioarts.com/press_release/ba09_09_09.htm (the URL still exists, but the content has changed—it no longer features Hawthorne's release); Woestendiek, *Dog, Inc.*

70*n* *The winner of the Golden Clone Giveaway*: "Golden Clone Giveaway," BioArts International, July 5, 2008, www.bestfriendsagain.com/goldenclonegiveaway/index.html (site discontinued, accessed via Internet Archive on June 24, 2011, http://web.archive.org/web/20080705173449/www.bestfriendsagain.com/goldenclonegiveaway/index.html).

70*n* *Team Trakr . . . trained to participate*: "Team Trakr," Team Trakr Foundation, accessed June 25, 2011, www.teamtrakr.org.

71 *Researchers sometimes inflict physical pain . . . stressful experiments*: For more on the difference between pain and suffering, and potential sources of both to lab animals, see ibid., 61–81.

71 *"Animals have the same desires . . ."*: Marc Bekoff, in discussion with author via telephone, November 2, 2011.

71 *In a 2008 report, the two groups*: The Humane Society of the United States and the American Anti-Vivisection Society, *Pet Cloning Is NOT for Pet Lovers* (May 22, 2008).

71 *"Few cloned animals . . ."*: Ibid.

71 *To create Snuppy, the South Korean . . . in an editorial*: "A Dog's Life," *Nature* 436 (August 4, 2005): 604.

71 *"It is unlikely that even . . ."*: Ibid.

72 *Animal Welfare Act*: Animal Welfare Act, 7 U.S.C., Chapter 54, Sections 2131–2159; "Animal Welfare," 9 C.F.R., Chapter 1, Subchapter A, Parts 1–3. You can read more about these regulations, and view the full text of the law, at "Animal Welfare Act," National Agricultural Library, United States Department of Agriculture, accessed March 26, 2012, http://awic.nal.usda.gov/government-and-professional-resources/federal-laws/animal-welfare-act.

72 *"positive physical contact . . . of the animal"*: "Exercise for Dogs," 9 C.F.R. Chapter 1, Subchapter A, Part 3, Section 3.8.

72 *Among other things, the code stated*: "Code of Bioethics," Genetic Savings & Clone, May 10, 2006, www.savingsandclone.com/ethics/code_of_bioethics_pet2000 .html (site discontinued, accessed via Internet Archive on February 7, 2012, http://web.archive.org/web/20060510145114/www.savingsandclone.com/ethics/code_of_bioethics_pet2000.html).

72*n* *Hawthorne countered some of these concerns*: Hawthorne, "A Project to Clone Companion Animals," 229–31.

72*n* *The A&M researchers got most of their cat eggs*: Mark Westhusin, e-mail to author, February 7, 2012.

72*n* *"never worked out well . . . for nuclear transfer"*: Westhusin, e-mail to author, February 2, 2012.

72*n* *Institutions that receive federal*: Office of Laboratory Animal Welfare, National Institutes of Health, *Public Health Service Policy on Humane Care and Use of Laboratory Animals* (rev. August 2002).

73 *Surveys show that more*: The Center for Genetics and Society keeps an excellent running tally of opinion polls on various types of animal cloning, "Animal and Pet Cloning Opinion Polls," Center for Genetics and Society, accessed June 2, 2011, www.geneticsandsociety.org/article.php?id=470. See also "Cloning," Gallup, Inc., accessed March 25, 2012 www.gallup.com/poll/6028/cloning.aspx; Lydia Saad, "Doctor-Assisted Suicide Is Moral Issue Dividing Americans Most," accessed on March 25, 2012, www.gallup.com/poll/147842/doctor-assisted-suicide-moral-issue-dividing-americans.aspx.

73 *Even the Animal Welfare Act*: Animal Welfare Act, 7 U.S.C., Chapter 54, Sections 2131–2159.

73 *the A&M team adhered*: Westhusin, e-mail message, February 2, 2012.

73 *Though cloners are figuring out*: Ibid.

74 *on September 10, 2009, he announced*: Lou Hawthorne, "Six Reasons We're No Longer Cloning Dogs."

74 *"Cloning," he acknowledged*: Ibid.

74 *Several cloned cows, for instance*: Walton, discussion, December 2010. "Apple 2 Wins at World Dairy Expo," Bovance, accessed June 24, 2011, www.bovance.com/news_102008.html.

74 *Doc, the winning steer*: Walton, discussion, December 2010; "Champion Steer at Iowa State Fair Continues Reign," Bovance, accessed June 24, 2011, www.bovance.com/news_083010.html.

74*n* *Dolly had been cloned again . . . just five embryos*: Fiona Macrae, "Dolly Reborn! Four Clones Created of Sheep That Changed Science," *Daily Mail*, November 30, 2010, www.dailymail.co.uk/sciencetech/article-1334201/Dolly-reborn-Four-clones-created-sheep-changed-science.html.

75 *$20,000 for the cow*: Walton, in discussion with author via telephone, February 6, 2012.

75 *more than pays for itself*: Walton, discussion, December 2010.

75 *cloning several hundred animals*: Walton, e-mail message to author, June 29, 2011.

75 *Most of its customers are duplicating cattle*: Ibid.

75 *ViaGen has cloned a champion barrel racer*: "Charmayne James and Scamper," ViaGen, accessed April 7, 2012, www.viagen.com/benefits/success-stories/horse-owners-and-breeders-james/.

75 *an Argentinian polo player*: Rory Carroll, "Argentinian Polo Readies Itself for Attack of the Clones," *Guardian*, June 5, 2011, www.guardian.co.uk/world/2011/jun/05/argentinian-polo-clones-player.

75 *And in 2012, the Fédération*: "FEI Spring Bureau Meeting Update," Fédération Equestre Internationale, June 18, 2012, accessed August 15, 2012, www.fei.org/media/press-releases/fei-spring-bureau-meeting-update. Kastalia Medrano, "Cloned Horses Coming to the Olympics?" *National Geographic News*, August 3, 2012, http://news.nationalgeographic.com/news/2012/08/120808-cloned-horses-clones-science-london-olympics-2012-equestrian/.

75*n* *Despite the furor over cloned meat . . . to be slaughtered*: Ibid.

75*n* *And although the FDA concluded*: Center for Veterinary Medicine, *Animal Cloning: A Risk Assessment*.

75*n* *Instead, cloned cows will be used . . . in grocery stores*: Walton, discussion, December 2010.

75*n* *ViaGen's financial backers*: Ibid.

76 *One of these companies, PerPETuate*: "About PerPETuate, Inc.," PerPETuate, Inc., accessed March 25, 2012, www.perpetuate.net/about/.

76 *"produce extraordinary, matchless . . ."*: Ibid.

76 *Gillespie guided me*: Information about PerPETuate comes from Ron Gillespie, in discussion with author via telephone: June 3, 2011.

77 *In the meantime, several of its customers*: Ibid.

77 *and Gillespie says the price*: Gillespie, discussion.

77 *"because it gives her hope"*: Ibid.

77 *"There is almost no fee . . ."*: "Our Pricing," PerPETuate, Inc., accessed March 25, 2012, www.perpetuate.net/pricing/.

77*n* *"theme park for cloned dogs"*: "RNL Bio and Start Licensing Settled Patent Disputes and Concluded a License Agreement," PR Newswire, www.prnewswire.com/news-releases/rnl-bio-and-start-licensing-settled-patent-disputes-and-concluded-a-license-agreement-81191552.html.

78 *$1,300, plus a yearly storage fee*: Ibid.

78 *All major credit cards*: Ibid.

78 *Westhusin, for instance, cloned a bull*: Kraemer, discussion, December 2010; Westhusin, discussion, May 2011; M. E. Westhusin et al., "Rescuing Valuable Genomes by Animal Cloning: A Case for Natural Disease Resistance in Cattle," *Journal of Animal Science* 85, no. 1 (2007): 138–42.

4. Nine Lives

80 *Almost a quarter . . . of birds*: Jean-Christophe Vié et al., eds., *Wildlife in a Changing World: An Analysis of the 2008 IUCN Red List of Threatened Species* (Gland, Switzerland: IUCN, 2009).

80 *There have been five*: Richard Leakey and Roger Lewin, *The Sixth Extinction: Patterns of Life and the Future of Humankind* (New York: Random House, 1996); Elizabeth Kolbert, "The Sixth Extinction," *The New Yorker*, May 25, 2009.

80 *demographers estimate that*: United Nations, Department of Economic and

Social Affairs, Population Division, *World Population Prospects: The 2010 Revision, Highlights and Advance Tables* (New York: United Nations, 2011).

81 *a little gaur named Noah*: Details about Noah and his cloning are from Robert P. Lanza et al., "Cloning of an Endangered Species (*Bos gaurus*) Using Interspecies Nuclear Transfer," *Cloning* 2, no. 2 (2000): 79–90; Philip Damiani et al., Cloning Endangered and Extinct Species, US Patent application no. 10/398,608, filed August 1, 2003, publication no. US 2004/0031069 A1.

81 *1,200 acres*: "Species Survival Center," Audubon Nature Institute, accessed April 20, 2012, www.auduboninstitute.org/conservation/saving-species /research-center.

82 *36,000-square-foot complex*: Ibid.

82 *Dresser has spent her life*: Information on Dresser's background comes from multiple conversations with her, including discussion with author via telephone, November 23, 2009; discussion with author, New Orleans, Louisiana, December 1, 2010; discussion with author via telephone, April 17, 2012.

83 *"I saw the science and technology . . ."*: Dresser, discussion, November 2009.

83 *At CREW, Dresser and her colleagues*: Dresser, discussion, December 2010; B. L. Dresser et al., "Induction of Ovulation and Successful Artificial Insemination in a Persian Leopard (*Panthera pardus saxicolor*)," *Zoo Biology* 1, no. 1 (1982): 55–57; C. E. Pope et al., "Birth of Western Lowland Gorilla (*Gorilla gorilla gorilla*) Following In Vitro Fertilization and Embryo Transfer," *American Journal of Primatology* 41, no. 3 (1997): 247–60.

83 *"We exist because we want . . ."*: Dresser, discussion, November 2009.

83 *whatever reproductive technologies*: For more on the use of assisted reproductive technologies with endangered species, see S. M. H Andrabi and W. M. C. Maxwell, "A Review on Reproductive Biotechnologies for Conservation of Endangered Mammalian Species," *Animal Reproduction Science* 99 (2007): 223–43.

84 *Nearly all of these felines . . . wild felines*: Dresser, discussions, November 2009 and December 2010; Martha C. Gomez et al., "Birth of African Wildcat Cloned Kittens Born from Domestic Cats," *Cloning and Stem Cells* 6 (2004): 247–58.

84 *the technique had limitations . . . even dead animals*: Dresser, discussion, November 2009; Dresser, in discussion with author via telephone, March 25, 2011; Robert P. Lanza et al., "Cloning Noah's Ark," *Scientific American*, November 2000, 84–89; William V. Holt et al., "Wildlife Conservation and Reproductive Cloning," *Reproduction* (March 2004): 317–24; Oliver A. Ryder, "Cloning Advances and Challenges for Conservation," *TRENDS in Biotechnology* 20, no. 6 (June 2002): 231–32.

84 *As she imagines it*: Dresser, discussions, December 2010 and April 2012.

85 *"If a habitat can't be saved . . ."*: Dresser, discussion, April 2012.

85 *"There is no one answer . . ."*: Dresser, discussion, December 2010.

85 *the African wildcat*: C. Driscoll and K. Nowell, "*Felis silvestris*," in *IUCN Red List of Threatened Species* 2011.2, www.iucnredlist.org/apps/redlist/details /8543/0; Dresser, discussion, December 2010.

86 *Nuclear transfer presents . . .* interspecies *nuclear transfer*: Dresser, discussions, November 2009 and March 2011; Lanza et al., "Cloning Noah's Ark"; Carrie Friese, "Enacting Conservation and Biomedicine: Cloning Animals of Endangered Species in the Borderlands of the United States" (dissertation, University of California, San Francisco, 2007).

86 *To clone Jazz*: The details of Jazz's cloning, and the birth of his clone, are from Gomez et al., "Birth of African Wildcat Cloned Kittens"; Dresser, discussion, March 2011.

86*n* *"[I]s a cloned animal . . ."*: David Ehrenfeld, "Transgenics and Vertebrate Cloning as Tools for Species Conservation," *Conservation Biology* 20, no. 3 (2006): 723–32.

87–88 *That November, Miles and Otis . . . various zoos*: "African Wildcat Clone Family Tree," Audubon Center for Research of Endangered Species, downloaded April 24, 2012, www.flickr.com/photos/audubonimages/3910008886/in/set-72157624320538361/; Dresser, discussions, November 2009 and April 2012.

88 *After these successes, the ACRES researchers . . . surrogate mothers*: Betsy Dresser, e-mail message to author, April 23, 2012.

88 *A European team made*: Pasqualino Loi et al., "Genetic Rescue of an Endangered Mammal by Cross-species Nuclear Transfer Using Post-mortem Somatic Cells," *Nature Biotechnology* 19 (October 2001): 962–64.

88 *Korean researchers cloned an endangered*: Min Kyu Kim et al., "Endangered Wolves Cloned from Adult Somatic Cells," *Cloning and Stem Cells* 9, no. 1 (2007): 130–37; H. J. Oh et al., "Cloning Endangered Gray Wolves (*Canis lupus*) from Somatic Cells Collected Postmortem," *Theriogenology* 70, no. 4 (2008): 638–47; Kim Tong-hyung, "Endangered Jeju Cattle Cloned," *Korea Times*, August 31, 2009, www.koreatimes.co.kr/www/news/tech/2012/03/129_51015.html.

88 *In 2012, scientists in India*: Aijaz Hussain, "Kashmir Scientists Clone Rare Cashmere Goat," Associated Press, March 15, 2012.

88 *After making Noah*: Constance Holden, "Banteng Cloned," *Science* (April 8, 2003): http://news.sciencemag.org/sciencenow/2003/04/08-01.html; "Cloned Endangered Species Euthanized," UPI, April 8, 2003, www.upi.com/Science_News/2003/04/08/Cloned-endangered-species-euthanized/UPI-42791049838441/.

89 *In Dresser's mind, the first task*: Dresser, discussions, November 2009 and April 2012.

89 *But researchers couldn't just*: Information about what's required to introduce captive-born animals into the wild is from Dresser, discussions, November 2009 and April 2012; International Union for Conservation of Nature and Natural Resources, *IUCN Guidelines for Re-introductions* (Gland, Switzerland: IUCN, 1998).

89 *from 11 to 53 percent*: K. R. Jule et al., "The Effects of Captive Experience in Reintroduction Survival in Carnivores: A Review and Analysis," *Biological Conservation* 141, no. 2 (2008): 355–63.

90 *black-footed ferrets*: J. Belant et al., "*Mustela nigripes*," *IUCN Red List of Threatened Species* 2011.2, www.iucnredlist.org/apps/redlist/details/14020/0.

90 *golden lion tamarins*: M. C. M. Kierulff et al., "*Leontopithecus rosalia*," in *IUCN Red List of Threatened Species* 2011.2, www.iucnredlist.org/apps/redlist /details/11506/0.

90 *Arabian oryx*: IUCN SSC Antelope Specialist Group 2011, "*Oryx leucoryx*," in *IUCN Red List of Threatened Species* 2011.2, www.iucnredlist.org/apps/redlist /details/15569/0.

90 *For example, some plants . . . of vegetation*: James A. Estes et al., "Trophic Downgrading of Planet Earth," *Science* 333 (July 15, 2011): 301–306; C. N. Johnson, "Ecological Consequences of Late Quaternary Extinctions of Megafauna," *Proceedings of the Royal Society B* 276 (2009): 2509–19.

90 *Today, it's a desolate place . . . these grasslands*: Sergey A. Zimov, "Pleistocene Park: Return of the Mammoth's Ecosystem," *Science* 308 (May 6, 2005): 796–98.

90 *"In the winter, the animals . . ."*: Ibid.

91 *Zimov is trying . . . shape the landscape*: Ibid.; see also "Pleistocene Park," Pleistocene Park, www.pleistocenepark.ru/en/.

91 *There are more radical proposals*: Josh Donlan, "Re-wilding North America," *Nature* 436 (August 18, 2005): 913–14; C. Josh Donlan et al., "Pleistocene Rewilding: An Optimistic Agenda for Twenty-First Century Conservation," *The American Naturalist* 168, no. 5 (2006): 660–81.

91 *Take the once-abundant gray wolf . . . thrive together*: William J. Ripple and Robert L. Beschta, "Trophic Cascades in Yellowstone: The First 15 Years After Wolf Reintroduction," *Biological Conservation* 145 (2012): 205–13; William J. Ripple and Robert L. Beschta, "Wolf Reintroduction, Predation Risk, and Cottonwood Recovery in Yellowstone National Park," *Forest Ecology and Management* 184 (2003): 299–313.

92–93 *such as the cheetah . . . of inbreeding*: Marilyn Menotti-Raymond and Stephen J. O'Brien, "Dating the Genetic Bottleneck of the African Cheetah," *Proceedings of the National Academy of Sciences USA* 90 (April 1993): 3172–76; S. Durant et al., "*Acinonyx jubatus*," in *IUCN Red List of Threatened Species* 2011.2, www.iucnredlist.org/apps/redlist/details/219/0.

93 *but we could use the technology*: How cloning could help maintain genetic diversity is from Dresser, discussions, November 2009 and March 2011; Lanza et al., "Cloning Noah's Ark"; Holt et al., "Wildlife Conservation and Reproductive Cloning"; and Ryder, "Cloning Advances and Challenges for Conservation."

93 *"That is years of science . . ."*: Dresser, discussion, December 2010.

93–94 *This is the Frozen Zoo . . . lions, tigers, and bears*: "Frozen Zoo," Audubon Nature Institute, accessed December 6, 2010, www.auduboninstitute.org/saving -species/frozen-zoo.

94 *"We should be systematically sampling . . ."*: Kraemer, discussion, October 2009.

94 *eighteen institutions in eight countries*: "Consortium," Frozen Ark Project, accessed April 9, 2012, www.frozenark.org/consortium.

94 *48,000 DNA samples from more than 5,500 species*: "Animals in the Ark," Frozen Ark Project, accessed April 9, 2012, www.frozenark.org/animals-ark.

95 *10,000 species by 2015*: "What We Need," Frozen Ark Project, accessed April 9, 2012, www.frozenark.org/what-we-need.

95 *For example, the San Diego Zoo's*: Andrea Johnson, "Preserving Hawaiian Bird Cell Lines," San Diego Zoo Global, November 7, 2008, http://blog.sandiegozooglobal.org/2008/11/07/preserving-hawaiian-bird-cell-lines/; BirdLife International 2009, "*Melamprosops phaeosoma*," in *IUCN Red List of Threatened Species* 2011.2, www.iucnredlist.org/apps/redlist/details/106008926/0; Mark Szotek, "Tales from a Frozen Zoo," Mongabay.com, February 2, 2010, http://news.mongabay.com/2010/0202-szotek_frozen_zoo.html.

95 *By 1999, there was only one . . . in other young clones*: Celia's demise, and subsequent cloning, are detailed in J. Folch et al., "First Birth of an Animal from an Extinct Subspecies (*Capra pyrenaica pyrenaica*) by Cloning," *Theriogenology* 71 (2009): 1026–34; Damiani et al., "Cloning Endangered and Extinct Species"; Richard Gray and Roger Dobson, "Extinct Ibex Is Resurrected by Cloning," *Telegraph*, January 31, 2009, www.telegraph.co.uk/science/science-news/4409958/Extinct-ibex-is-resurrected-by-cloning.html.

96 *Australian researchers want to revive*: Deborah Smith, "Tassie Tiger Cloning 'Pie-in-the-Sky Science,'" *Sydney Morning Herald*, February 17, 2005, www.smh.com.au/news/science/tassie-tiger-cloning-pieinthesky-science/2005/02/16/1108500157295.html; Daniel Dasey, "Researchers Revive Plan to Clone the Tassie Tiger," *Sydney Morning Herald*, May 15, 2005, www.smh.com.au/news/Science/Clone-again/2005/05/14/1116024405941.html; "The Thylacine: A Case Study," Biotechnology Online, Commonwealth Scientific and Industrial Research Organisation, accessed April 20, 2012, www.biotechnologyonline.gov.au/enviro/thylacine.html.

96 *Like a kangaroo . . . Tasmanian tiger*: "*Thylacinus cynocephalus*—Thylacine," Department of Sustainability, Environment, Water, Population and Communities, Commonwealth of Australia, accessed April 21, 2012, www.environment.gov.au/cgi-bin/sprat/public/publicspecies.pl?taxon_id=342.

96 *The mammal has been extinct*: Andrew J. Pask et al., "Resurrection of DNA Function *In Vivo* from an Extinct Genome," *PLoS One* 3, no. 5 (2008): e2240.

96 *In 2008, one team of scientists*: Ibid.; Katherine Sanderson, "Tasmanian Tiger Gene Lives Again," *Nature* (May 20, 2008): www.nature.com/news/2008/080520/full/news.2008.841.html.

96*n* *One dream is to resurrect*: Henry Nicholls, "The Legacy of Lonesome George," *Nature*, July 18, 2012, www.nature.com/news/the-legacy-of-lonesome-george-1.11017.

97 *"[W]e have restored to life . . ."*: Pask et al., "Resurrection of DNA Function."

97 *The following year, a different group*: Webb Miller et al., "The Mitochondrial

Genome Sequence of the Tasmanian Tiger (*Thylacinus cynocephalus*)," *Genome Research* 19, no. 2 (2009): 213–20.

97 *wombats and wallabies*: "*Thylacinus cynocephalus*," Department of Sustainability.

97 *Russian, Japanese, and Korean*: Shingo Ito, "Researchers Aim to Resurrect Mammoth in Five Years," AFP, January 17, 2011; "S. Korean, Russian Scientists Bid to Clone Mammoth," AFP, March 12, 2012; Jaeyeon Woo, "Will Resurrecting a Mammoth Be Possible?" *Wall Street Journal*, March 13, 2012, http://blogs.wsj.com/korearealtime/2012/03/13/will-be-resurrecting-a-mammoth-possible/.

97 *During his 2006 trial*: Reuters, "Korean Scientist Paid Russia Mafia for Mammoth," NBCNEWS.com, October 24, 2006, www.msnbc.msn.com/id/15399222/ns/technology_and_science-science/t/korea-scientist-paid-russia-mafia-mammoth/#.UCrCPkTTMgq.

97 *That will be a difficult task*: Henry Nicholls, "Let's Make a Mammoth," *Nature* 456 (November 20, 2008): 311.

97 *the DNA in even the best*: Evgeny I. Rogaev et al., "Complete Mitochondrial Genome and Phylogeny of Pleistocene Mammoth *Mammuthus primigenius*," *PLoS Biology* 4, no. 3 (2006): e73.

97 *The other option*: Nicholls, "Let's Make a Mammoth."

98 *Among other obstacles*: Ibid., 312.

98 *famed paleontologist Jack Horner*: Jack Horner and James Gorman, "Dinosaur Resurrection," *Discover*, April 2009, 50–53. See also Jack Horner and James Gorman, *How to Build a Dinosaur* (New York: Penguin, 2009).

99 *To many biologists, cloning*: Holt et al., "Wildlife Conservation and Reproductive Cloning"; Friese, "Enacting Conservation and Biomedicine"; Dresser, discussion, December 2010; Ehrenfeld, "Transgenics and Vertebrate Cloning."

99 *"is a glamorous technology . . ."*: Ehrenfeld, "Transgenics and Vertebrate Cloning."

99 *Cloning could help us maintain species*: Andrabi and Maxwell, "A Review on Reproductive Biotechnologies."

100 *Scientists have managed to take*: Inbar Friedrich Ben-Nun et al., "Induced Pluripotent Stem Cells from Highly Endangered Species," *Nature Methods* 8 (2011): 829–31; Ewen Callaway, "Could Stem Cells Rescue an Endangered Species?" *Nature*, September 4, 2011, www.nature.com/news/2011/110904/full/news.2011.517.html.

100 *"I may not live . . ."*: Dresser, discussion, December 2010.

100 *After spending years*: Details of Dresser's activities since leaving ACRES are from Dresser, discussions, March 2011 and April 2012. .

5. Sentient Sensors

102 *In the idyllic decades*: Information about grizzly bears in Yellowstone, human-bear conflicts, and the Craigheads' tracking project is from Mark A. Haroldson

et al., "Grizzly Bears in the Greater Yellowstone Ecosystem: From Garbage, Controversy, and Decline to Recovery," *Yellowstone Science* 16, no. 2 (2008): 13–22; Etienne Benson, *Wired Wilderness: Technologies of Tracking and the Making of Modern Wildlife* (Baltimore, MD: Johns Hopkins University Press, 2010); Etienne Benson, in discussion with author via telephone, September 30, 2011.

103 *"Beep, beep, beep . . ."*: This quote first appeared in Frank Craighead's book *Track of the Grizzly*, published in 1979. I came across it in Benson's *Wired Wilderness* (pp. 60–61).

104 *during the 1960s and '70s*: Gerald L. Kooyman, "Genesis and Evolution of Biologging Devices: 1963–2002," *Memoirs of National Institute of Polar Research* 58 (2004): 15–22.

104 *one scientist measured*: Ibid.

104 *But biologists and engineers stuck with it*: For a history and evolution of dive loggers see ibid.

104 *some smaller than a jelly bean*: "Miniaturized Wildlife Tracking Tags Deployed Worldwide Collect Crucial Data," Atlantic Canada Opportunities Agency, accessed April 14, 2012, www.acoa-apeca.gc.ca/eng/ImLookingFor/ProgramInformation/AtlanticInnovationFund/Pages/LotekWirelessInc.aspx?ProgramID.

104 *Heavy fishing, pollution*: For more on threats to oceans and marine life, see R. A. Myers and C. A. Ottensmeyer, "Extinction Risk in Marine Species," in *Marine Conservation Biology: The Science of Maintaining the Sea's Biodiversity*, ed. E. A. Norse and L. B. Crowder (Washington, DC: Island Press, 2005).

104 *Populations of marine animals*: Boris Worm et al., "The Future of Marine Animal Populations," in *Life in the World's Oceans: Diversity, Distribution, and Abundance*, ed. Alasdair D. McIntyre (UK: Wiley-Blackwell, 2010), 315–30.

105 *Between 2000 and 2009 . . . species converge*: B. A. Block et al., "Tracking Apex Marine Predator Movements in a Dynamic Ocean," *Nature* 475 (July 7, 2011): 86–90.

105 *"When we start to understand . . ."*: Randy Kochevar, in discussion with author, Monterey, California, March 30, 2011.

105*n* *TOPP was one of seventeen . . . deepwater vents*: Jesse H. Ausubel et al., eds., *First Census of Marine Life 2010: Highlights of a Decade of Discovery* (Washington, DC: Census of Marine Life, 2010).

106 *150,000 gallons of seawater*: "The Facility," Tuna Research and Conservation Center, accessed March 6, 2012, www.tunaresearch.org/about/history.html.

106 *In 2012, a 593-pound specimen*: "Japan Tuna Sale Smashes Record," BBC News, January 5, 2012, www.bbc.co.uk/news/world-asia-pacific-16421231.

106 *bluefin can live for thirty years*: Kochevar, discussion, March 2011.

106 *grow to be thirteen feet . . . 45 miles per hour*: "Atlantic Bluefin Tuna (*Thunnus thynnus*)," NOAA Fisheries, accessed April 12, 2012, www.nmfs.noaa.gov/pr/species/fish/bluefintuna.htm.

106 *from South America to Norway*: Gareth L. Lawson, "Movements and Diving Behavior of Atlantic Bluefin Tuna *Thunnus thynnus* in Relation to Water Col-

umn Structure in the Northwestern Atlantic," *Marine Ecology Progress Series* 400 (February 11, 2012): 245–65.

106 *Bluefin tuna swim so fast*: Barbara A. Block, "Archival Tagging of Atlantic Bluefin Tuna (*Thunnus thynnus thynnus*)," *Marine Technology Society Journal* 32, no. 1 (1998): 37–46.

106–107 *Marine biologists use . . . reach of satellites*: Barbara Block, "Physiological Ecology in the 21st Century: Advancements in Biologging Science," *Integrative & Comparative Biology* 45 (2005): 305–20; Kochevar, discussion, March 2011.

107 *they realized that they could . . . reconstruct the tuna's path*: John Gunn and Barbara Block, "Advances in Acoustic, Archival, and Satellite Tagging of Tunas," in *Tuna: Physiology, Ecology, and Evolution*, ed. Barbara A. Block and E. Donald Stevens (San Diego, CA: Academic Press, 2001), 178–79; Block, "Archival Tagging of Atlantic Bluefin Tuna"; Kochevar, discussion, March 2011.

107 *To deploy the tags*: Details on the tagging procedure come from A. M. Boustany et al., "Movements of Pacific Bluefin Tuna (*Thunnus orientalis*) in the Eastern North Pacific Revealed with Archival Tags," *Progress in Oceanography* 86 (2010): 94–104; Block, "Archival Tagging of Atlantic Bluefin Tuna"; Barbara A. Block and Shana Miller, "Unveiling the Secret Life of an Ocean Giant," in *World Record Game Fishes* (International Game Fish Association, 2007), 84–92. The researchers also blog about their experiences out on the ocean. To read firsthand accounts of catching and tagging tuna, check out http://trccblog.blogspot.com and http://gtopp.blogspot.com/.

107 *The tag is a marvel*: The tag specs cited pertain specifically to the Lotek LTD 2310 archival tag. See "Lotek Archival Tag Series," Lotek Wireless, downloaded March 26, 2012, www.lotek.com/lat-geo-ext-mem.pdf; "Locating Tuna in the Open Ocean," Lotek Wireless, downloaded April 14, 2012, www.lotek.com/locatingtuna.pdf. This particular model is commonly used, including by Block and her colleagues. (For instance, they used this model in the research published as Boustany et al., "Movements of Pacific Bluefin Tuna (*Thunnus orientalis*) in the Eastern North Pacific.")

107*n* *The TOPP team, for instance . . . switches off*: Block, "Physiological Ecology in the 21st Century"; Kochevar, discussion, March 2011; "About TOPP," Tagging of Pacific Predators, accessed October 19, 2010, www.topp.org/about_topp.

108 *The scientists also attach . . . streamer tag says*: "Found an electronic tag? Contact us today, and claim your reward," Tag-A-Giant Foundation, accessed April 11, 2012, http://tagagiant.org/Reward.shtml.)

108 *They use a combination of readings*: Kochevar, discussion, March 2011.

109 *"You think of it as this . . ."*: Alex Norton, in discussion with author, Monterey, California, March 30, 2011.

110 *In the 1960s . . . irritation, and pain*: Benson, *Wired Wilderness*; Benson, discussion.

110 *In some studies, for example*: Cindy L. Hull, "The Effect of Carrying Devices on Breeding Royal Penguins," *The Condor* 99, no. 2 (1997): 530–34; Sabrina S.

Taylor et al., "Foraging Trip Duration Increases for Humboldt Penguins Tagged with Recording Devices," *Journal of Avian Biology* 32, no. 4 (2001): 369–72; Donald A. Croll et al., "Foraging Behavior and Reproductive Success in Chinstrap Penguins: The Effects of Transmitter Attachment," *Journal of Field Ornithology* 67, no. 1 (1996): 1–9.

110 *In certain species of fish*: C. J. Bridger and R. K. Booth, "The Effects of Biotelemetry Transmitter Presence and Attachment Procedures on Fish Physiology and Behavior," *Reviews in Fisheries Science* 11, no. 1 (2003): 13–34.

110–11 *Surgically implanted tags . . . and pathogens*: Many researchers have written about ways in which tags could affect wild animals. A few of the sources I consulted are Rory P. Wilson and Clive R. McMahon, "Measuring Devices on Wild Animals: What Constitutes Acceptable Practice?" *Frontiers in Ecology and the Environment* 4, no. 3 (2006): 147–54; Penny Hawkins, "Biologging and Animal Welfare: Practical Refinements," *Memoirs of the National Institute for Polar Research* 58 (2004): 58–68; Roger A. Powell and Gilbert Proulx, "Trapping and Marking Terrestrial Mammals for Research: Integrating Ethics, Performance Criteria, Techniques, and Common Sense," *ILAR Journal* 44, no. 3 (2003): 259–76; R. J. Putman, "Ethical Considerations and Animal Welfare in Ecological Field Studies," *Biodiversity and Conservation* 4 (1995): 903–15; Russell J. Borski and Ronald G. Hodson, "Fish Research and the Institutional Animal Care and Use Committee," *ILAR Journal* 44, no. 4 (2003): 286–94; American Fisheries Society, *Guidelines for the Use of Fishes in Research* (Bethesda, MD: 2004), available at www.fisheries.org/afs/docs/policy_16.pdf; Robert S. Sikes et al., "Guidelines of the American Society of Mammalogists for the Use of Wild Mammals in Research," *Journal of Mammalogy* 92, no. 1 (2011): 235–53.

111 *It's not easy to perform*: Kochevar, discussion, March 2011.

111 *So Block and her colleagues . . . from their bodies*: Block, "Archival Tagging of Atlantic Bluefin Tuna"; Block and Miller, "Unveiling the Secret Life of an Ocean Giant"; Kochevar, discussion, March 2011.

111 *classified as "endangered" or "threatened"*: "Loggerhead Turtle (*Caretta caretta*)," NOAA Fisheries, accessed April 11, 2012, www.nmfs.noaa.gov/pr/species/turtles/loggerhead.htm.

111 *The loggerheads that live*: Jeffrey J. Polovina et al., "Turtles on the Edge: Movement of Loggerhead Turtles (*Caretta caretta*) Along Oceanic Fronts, Spanning Longline Fishing Grounds in the Central North Pacific, 1997–1998," *Fisheries Oceanography* 9, no. 1 (2000): 71–82.

112 *Federal regulations stipulate . . . until the next year*: Jeffrey Polovina, in discussion with author via telephone, November 3, 2011.

112 *Polovina, Howell, and their colleagues . . . hungry turtles*: Polovina et al., "Turtles on the Edge"; J. J. Polovina et al., "The Transition Zone Chlorophyll Front, a Dynamic Global Feature Defining Migration and Forage Habitat for Marine Resources," *Progress in Oceanography* 49 (2001): 469–83; J. J. Polovina et al., "Forage and Migration Habitats of Loggerhead (*Caretta caretta*) and Olive

Ridley (*Lepidochelys olivacea*) Sea Turtles in the Central North Pacific Ocean," *Fish Oceanography* 13 (2004): 36–51; Evan A. Howell et al., "TurtleWatch: A Tool to Aid in the Bycatch Reduction of Loggerhead Turtles *Caretta caretta* in the Hawaii-Based Pelagic Longline Fishery," *Endangered Species Research* 5, no. 2–3 (2008): 267–78; Polovina, discussion.

112 *Since December 2006, that's what*: Details about the program are from Polovina, discussion; Howell et al., "TurtleWatch"; "EOD TurtleWatch," NOAA Pacific Islands Fisheries Science Center, accessed March 12, 2012, www.pifsc.noaa.gov/eod/turtlewatch.php.

113 *Since the program began*: Polovina, discussion.

113 *the overexploited tuna populations*: Barbara A. Block et al., "Migratory Movements, Depth Preferences, and Thermal Biology of Atlantic Bluefin Tuna," *Science* 293 (August 17, 2001): 1310–14.

113 *Since the early 1980s*: Lawson, "Movements and Diving Behavior of Atlantic Bluefin Tuna."

113 *International Commission for the Conservation of Atlantic Tunas*: Information about how ICCAT manages Atlantic bluefin tuna is from Block et al., "Migratory Movements, Depth Preferences, and Thermal Biology"; Barbara A. Block et al., "Electronic Tagging and Population Structure of Atlantic Bluefin Tuna," *Nature* 434 (April 28, 2005): 1121–27; Charles H. Greene et al., "Advances in Conservation Oceanography: New Tagging and Tracking Technologies and Their Potential for Transforming the Science Underlying Fisheries Management," *Oceanography* 22, no. 1 (2009): 210–23; Kochevar, discussion, March 2011.

113 *more than 90 percent since 1970*: Andreas Walli et al., "Seasonal Movements, Aggregations and Diving Behavior of Atlantic Bluefin Tuna (*Thunnus thynnus*) Revealed with Archival Tags," *PLoS One* 4, no. 7 (2009).

113 *smaller than the eastern one . . . more stringent*: Greene et al., "Advances in Conservation Oceanography"; Randy Kochevar, in discussion with author via telephone, October 31, 2011.

113 *"Well, when we started tagging . . ."*: Kochevar, discussion, March 2011.

113 *Their tracking data reveals*: Block et al., "Migratory Movements, Depth Preferences, and Thermal Biology"; Block et al., "Electronic Tagging and Population Structure"; Walli et al., "Seasonal Movements, Aggregations and Diving Behavior"; Greene et al., "Advances in Conservation Oceanography"; Kochevar, discussions, March and October 2011; Steve Teo, in discussion with author via telephone, October 21, 2011.

114 *and points the way*: Kochevar, discussion, March 2011; Teo, discussion; Block and Miller, "Unveiling the Secret Life of an Ocean Giant."

114 *Block's team, for instance*: Mark Shwartz and Ken Peterson, "Study: Better Protections for Bluefin Tuna Needed," Stanford News Service, April 27, 2005, http://news.stanford.edu/news/2005/may4/tuna-042705.html; Andrew C. Revkin, "Tracking the Imperiled Bluefin from Ocean to Sushi Platter," *New York Times*, May 3, 2005, www.nytimes.com/2005/05/03/science/earth/03tuna.html?pagewanted=all.

114 *spend their lives in one*: Information about elephant seal habitat, and scientists' difficulty getting to it, is from Michael Fedak, in discussion with author via telephone, November 17, 2011; and J. Charrassin et al., "Southern Ocean Frontal Structure and Sea-ice Formation Rates Revealed by Elephant Seals," *PNAS* 10, no. 33 (2008): 11634–39.

114 *more than a mile*: J. Charrassin et al., "New Insights into Southern Ocean Physical and Biological Processes Revealed by Instrumented Elephant Seals," in *Proceedings of OceanObs '09: Sustained Ocean Observations and Information for Society* 2 (Venice, September 21–25, 2009).

114 *"they might as well be going off . . ."*: Fedak, discussion.

114 *Eager to learn more*: Ibid.

114–15 *Between 2003 and 2007 . . . 102 elephant seals*: Charrassin et al., "New Insights into Southern Ocean."

114*n* *As of 2010, 28 percent . . . 53 percent*: Exploitation percentages come from Fisheries and Aquaculture Department, Food and Agriculture Organization of the United Nations, *The State of World Fisheries and Aquaculture* (Rome: FAO, 2010), 8.

115 *Whenever an elephant seal . . . back to the lab*: Ibid.; Fedak, discussion.

115 *"except stuck on something hairy . . ."*: Fedak, discussion.

115 *As the numbers started*: Ibid.

115 *"These guys needed this data . . ."*: Ibid.

115 *Oceanographers are now using*: Ibid.

115 *Among other things, tagged elephant seals . . . faster than expected*: Laurie Padman et al., "Seals Map Bathymetry of the Antarctic Continental Shelf," *Geophysical Research Letters* 37, no. 21 (2010): 1–5; Donna Hesterman, "Elephant Seals Improve Maps of Antarctic Seafloor," University of California, Santa Cruz, October 15, 2010, http://news.ucsc.edu/2010/10/seal-maps.html.

115 *Today, marine mammals*: M. A. Fedak, "The Impact of Animal Platforms on Polar Ocean Observation" (paper under review); Michael Fedak, e-mail message to author, April 3, 2012.

115 *the U.S. Integrated Ocean Observing System*: National Oceanic and Atmospheric Administration, "National Ocean Observing System to See Marine Animal Migration, Adaptation Strategies," March 4, 2011, available at http://gtopp.org/images/stories/press_releases/03-04-11_Marine_Tagging.pdf; Kochevar, discussion, March 2011.

115 *Ice melt is just the beginning*: For more on the effects of climate change on the ocean and the organisms that live there, see J. A. Learmonth et al., "Potential Effects of Climate Change on Marine Mammals," *Oceanography and Marine Biology: An Annual Review* (2006): 431–64; Christopher D. G. Harley et al., "The Impacts of Climate Change in Coastal Marine Systems," *Ecology Letters* 9, no. 2 (2006): 228–41; R. Schubert et al., *The Future Oceans—Warming Up, Rising High, Turning Sour* (Berlin: German Advisory Council on Global Change, 2006); FAO Fisheries and Aquaculture Department, *The State of World Fisher-*

ies, 115–120; K. Cochrane et al., eds., *Climate Change Implications for Fisheries and Aquaculture: Overview of Current Scientific Knowledge* (Rome: FAO Fisheries and Aquaculture Department, 2009).

115 *These shifts are already*: Camille Parmesan, "Ecological and Evolutionary Responses to Recent Climate Change," *Annual Review of Ecology, Evolution, and Systematics* 37 (2006): 637–69.

115 *As waters warm*: Allison L. Perry et al., "Climate Change and Distribution Shifts in Marine Fishes," *Science* 308 (June 24, 2005): 1912–15.

116 *there have been shifts . . . fewer calves*: Learmonth et al., "Potential Effects of Climate Change."

116 *For example, scientists have used tags . . . conditions are like there*: M. Biuw et al., "Blubber and Buoyancy: Monitoring the Body Condition of Free-Ranging Seals Using Simple Dive Characteristics," *Journal of Experimental Biology* 206 (2003): 3405–23; Fedak, discussion.

116 *"We can then run models . . . happen to the beasts?'"*: Fedak, discussion.

116 *"We're making colleagues . . ."*: Ibid. See alsoMike Fedak, "Marine Animals as Platforms for Oceanographic Sampling: A 'Win/Win' Situation for Biology and Operational Oceanography," *Memoirs of the National Institute for Polar Research* 58 (2004): 133–47.

117 *Ocean Tracking Network*: Details on the Ocean Tracking Network are from *An Evolution in Ocean Research*, downloaded April 12, 2012, http://oceantrackingnetwork.org/images/brochure.pdf; Ocean Tracking Network, *Annual Report 2010–2011*, available at http://oceantrackingnetwork.org/AR_2010-2011.pdf; "Ocean Tracking Network," Dalhousie University, accessed April 2, 2012, http://oceantrackingnetwork.org/; "About the Project," Dalhousie University, accessed April 2, 2012, http://oceantrackingnetwork.org/aboutproject/index.html; "Ocean Monitoring," Dalhousie University, accessed April 2, 2012, http://oceantrackingnetwork.org/aboutproject/ocean.html; "Underwater Innovation: Canadian Technology at the Forefront," Dalhousie University, accessed April 2, 2012, http://oceantrackingnetwork.org/aboutproject/technology.html; "OTN South Africa Phase I Deployments Complete," Dalhousie University, accessed April 12, 2012, http://oceantrackingnetwork.org/news/safdeploy.html; "Halifax Line Is Now OTN's Longest Listening Line," Dalhousie University, accessed April 12, 2012, http://oceantrackingnetwork.org/news/hfx166.html.

117 *A team of Hawaiian biologists*: Kim N. Holland et al., "Inter-animal Telemetry: Results from First Deployment of Acoustic 'Business Card' Tags," *Endangered Species Research* 10 (2009): 287–93.

117 *A number of other labs*: Emily L. C. Shepard et al., "Identification of Animal Movement Patterns Using Tri-axial Accelerometry," *Endangered Species Research* 10 (2010): 47–60; Nicholas M. Whitney et al., "Identifying Shark Mating Behaviour Using Three-Dimensional Acceleration Loggers," *Endangered Species Research* 10 (2010): 71–82; John P. Skinner et al., "Head Striking During

Fish Capture Attempts by Steller Sea Lions and the Potential for Using Head Surge Acceleration to Predict Feeding Behavior," *Endangered Species Research* 10 (2010): 61–69.

117 *"pop-up" satellite tag*: Block, "Physiological Ecology in the 21st Century," 308; Kochevar, discussion, March 2011.

117–18 *These tags are bigger, heavier*: John Gunn, "From Plastic Darts to Pop-up Satellite Tags," in *Fish Movement and Migration,* ed. D. A. Hancock et al. (Sydney: Australian Society for Fish Biology, 2000), 55–60. Additional price and size information came from comparing archival and pop-off products offered by Wildlife Computers, a leading tag manufacturer. (See www.wildlifecomputers .com/default.aspx.)

118 *Block, who piloted the use*: Barbara A. Block et al., "A New Satellite Technology for Tracking the Movements of Atlantic Bluefin Tuna," *Proceedings of the National Academy of Sciences USA* 95 (August 1998): 9384–89.

118 *A Canadian company*: "NanoTag Series Coded Radio Transmitters," Lotek Wireless, Inc., accessed April 14, 2012, www.lotek.com/nanotag.htm.

118 *In 2010, researchers reported*: M. Wikelski et al., "Large-Range Movements of Neotropical Orchid Bees Observed via Radio Telemetry," *PLoS One* 5, no. 5 (2010): e10738.

118 *a group of Swedish scientists*: Mercy Lard, et al., "Tracking the Small with the Smallest—Using Nanotechnology in Tracking Zooplankton," *PLoS ONE* 5, no. 10 (2010): e13516.

118 *"We're tracking everything . . . All the time"*: Benson, discussion.

118 *"Wildlife managers needed . . ."*: Ibid.

119 *"Do we really think . . ."*: Ibid.

119 *TOPP researchers, for instance*: "TOPP," Tagging of Pacific Predators, accessed April 3, 2012, http://topp.org/.

119 *a loser male . . . named Jonathan Sealwart*: "Jon Sealwart Everybody!! Today, Can Loser Males Have Hope Too??" Tagging of Pacific Predators, accessed April 3, 2012, www.topp.org/blog/jon_sealwart_everybody_today_can_loser _males_have_hope_too.

120 *"You see conservation organizations . . ."*: Benson, discussion.

120 *"[G]iving an animal a name . . ."*: S. Borkfelt, "What's in a Name?—Consequences of Naming Non-Human Animals," *Animals* 1, no. 1 (2011): 116–25.

121 *more than five hundred of them*: As of April 14, 2012.

6. Pin the Tail on the Dolphin

122 *In December 2005*: Information on Winter's accident and rescue comes from several sources, including multiple interviews with Kevin Carroll, exhibits and presentations at the Clearwater Marine Aquarium, and a documentary about Winter produced by the aquarium (*Winter . . . The Dolphin That Could,* produced and directed by David Yates and Steve Brown [Clearwater, FL: Clearwater Marine Aquarium, 2010], DVD).

124 *She's adapted to her . . . curve unnaturally*: Kevin Carroll, in discussion with author via telephone, June 8, 2010, and March 30, 2012.

124 *In September 2006, an aquarium*: Dana Zucker, interview by Melissa Block, *All Things Considered*, NPR, September 26, 2006, transcript available at www.npr.org/templates/story/story.php?storyId=6147502.

124 *A prosthetist named Kevin . . . lifetime of disability*: Carroll, discussions, June 2010 and March 2012.

124 *Carroll grew up near*: Information on Kevin Carroll's background and experience comes from numerous conversations with him between 2010 and 2012.

125 *"whatever comes our way"*: Carroll, discussion, June 2010.

125 *"I've sort of become the Dr. Doolittle . . ."*: Carroll, discussion, March 2012.

125 *Strzempka, who has worn . . . as a patient*: Kevin Carroll and Dan Strzempka, in discussion with author, Clearwater, Florida, March 25, 2011.

125 *"At first, I thought he was . . ."*: Dan Strzempka, in discussion with author, Clearwater, Florida, March 25, 2011.

126 *Carroll and Strzempka began a crash course*: Carroll and Strzempka, discussion, March 2011.

126–27 *In Winter's case, the basic plan . . . to the ground*: Ibid.

127 *"It was a small warehouse"*: Strzempka, discussion, March 2011.

127 *the engineer nailed it*: ALPS, the company that manufactures the new gel, would not reveal the exact composition of the material, saying that the formula is proprietary. But Strzempka says that possible tweaks include changing the ratio of plastics used or selecting new binding agents—compounds used to bulk up the mixture and hold it together. Materials scientists can also change the properties of a thermoplastic by playing with specifics of the curing process, during which heat and pressure are used to turn the liquid concoction into a durable solid, he said.

127 *"It's incredible material"*: Carroll, discussion, March 2011.

127 *Technically, the material*: Strzempka, discussion, March 2011.

127–28 *To make sure that Winter . . . heavier devices*: Carroll, discussion, June 2010.

128 *To put the device on . . . apparatus on*: Carroll and Strzempka, discussion, March 2011; Carroll, discussion, June 2010.

128 *Winter has to be supervised*: Strzempka, discussion, March 2011.

128 *Instead, it's reserved . . . from side to side*: Carroll and Strzempka, discussion, March 2011.

128 *"It's just beautiful . . ."*: Carroll, discussion, June 2010.

129 *Winter's scoliosis has improved*: Carroll and Strzempka, discussion, March 2011; Carroll, discussion, March 2012.

129 *Carroll and Strzempka are still*: Carroll, discussion, June 2010; Carroll and Strzempka, discussion, March 2011.

129 *For instance, Strzempka says*: Strzempka, discussion, March 2011.

129 *"mad scientist character"*: Kevin Carroll, in discussion with author, Stratford, Connecticut, October 5, 2010.

130 *Children with artificial arms*: Carroll, discussion, June 2010.

130 *"The psychological aspect . . ."*: Ibid.

130 *as word spread*: Carroll, discussion, October 2010; Strzempka, discussion, March 2011.

130 *"The stickiness is . . ."*: Strzempka, discussion, March 2011.

130 *It didn't take long*: "The Never Ending Tale of WintersGel," Hanger Prosthetics and Orthotics, accessed March 17, 2012, www.hanger.com/prosthetics/experience/patientprofiles/winterthedolphin/Pages/WintersGel.aspx.

130 *seasoned triathletes*: Dan Strzempka, in discussion with author via telephone, June 14, 2011.

130 *eleven-year-old girls*: "Megan McKeon," Hanger Prosthetics and Orthotics, accessed March 17, 2012, www.hanger.com/prosthetics/experience/patientprofiles/winterthedolphin/Pages/MeganMcKeon.aspx.

130 *"Animals give back . . ."*: Carroll, discussion, March 2012.

130 *some neuroscience research has*: Massimo Filippi et al., "The Brain Functional Networks Associated to Human and Animal Suffering Differ Among Omnivores, Vegetarians and Vegans," *PLoS One* 5, no. 5 (2010). Another study revealed that we have the most empathy for species that are most closely related to humans: H. Rae Westbury and David L. Neumann, "Empathy-Related Responses to Moving Film Stimuli Depicting Human and Non-human Animal Targets in Negative Circumstances," *Biological Psychology* 78, no. 1 (2008): 66–74.

129–30 *Animals have all sorts . . . roll their eyes*: Committee on Recognition and Alleviation of Pain in Laboratory Animals, National Research Council, *Recognition and Alleviation of Pain in Laboratory Animals* (Washington, DC: National Academies Press, 2009), 50.

131 *Mice make grimaces*: Dale J. Langford et al., "Coding of Facial Expressions of Pain in the Laboratory Mouse," *Nature Methods* 7 (May 9, 2010): 447–49.

131 *Carroll ultimately decided to pass*: Kevin Carroll, in discussion with author via telephone, February 8, 2011.

132 *Take the dog owner Gregg Miller*: Information about Buck's neutering, and the origins of Neuticles, are from Gregg Miller, in discussion with author via telephone, June 30, 2010.

132 *"Good God, it was . . . losing a body part?"*: Ibid.

132 *"People thought I was nuts"*: Ibid.

132 *Working with veterinarians*: "Neuticles Inventor Gregg A. Miller," CTI Neuticles, accessed November 2, 2011, www.neuticles.com/.

132 *"texture and firmness"*: "Interesting Facts About Neuticles," CTI Neuticles, accessed March 20, 2012, www.neuticles.com/facts.php.

132 *The first dog received*: "Neuticles Inventor Gregg A. Miller," CTI Neuticles.

132 *The Neuticles were popped in . . . surgical procedure*: "Frequently Asked Questions," CTI Neuticles, accessed November 2, 2011, www.neuticles.com/faq.php.

132 *prices range from*: Neuticles Brochure, CTI Neuticles, downloaded March 19, 2012, from www.neuticles.com/NeuticlesBrochure.pdf.

132 *The company also sells*: Neuticles Brochure, CTI Neuticles.

132*n* *"Even though he never got Neuticles . . ."*: Miller, discussion.

133 *250,000 pets in forty-nine countries*: "Interesting Facts About Neuticles," CTI Neuticles.

133 *"It will not help . . . inadequate to the task"*: T. Nagel, "What Is It Like to Be a Bat?" *The Philosophical Review* 83, no. 4 (1974): 435–50.

133*n* *Gregory Berns, a neuroscientist*: Gregory S. Berns et al., "Functional MRI in Awake Unrestrained Dogs," *PLoS One* 7, no. 5 (2012): e38027.

134 *can cause behavioral changes*: M. V. Kustritz, "Determining the Optimal Age for Gonadectomy of Dogs and Cats," *Journal of the American Veterinary Medical Association* 231, no. 11 (2007): 1665–75.

134 *"Pets don't have . . ."*: "Myths and Facts About Spaying and Neutering," Humane Society of the United States, accessed March 12, 2012, www.humanesociety.org/issues/pet_overpopulation/facts/spay_neuter_myths_facts.html.

134 *"a creepy, creepy thing . . ."*: Miller, discussion.

134 *Scientists studying the effects*: A. Brent Richards et al., "Gonadectomy Negatively Impacts Social Behavior of Adolescent Male Primates," *Hormones and Behavior* 56, no. 1 (2009): 140–48.

134 *"freaking out over . . ."*: Miller, discussion.

135 *some animal welfare groups*: These groups include the ASPCA, Canadian Humane Society, and New York Humane Society, as per www.neuticles.com/facts.php.

135 *In fact, when Miller heard*: "'Buddy Needs Neuticles' Proposal Reviewed by Clinton," CTI Neuticles, accessed November 2, 2011, www.neuticles.com/press.php.

135 *According to a survey of nearly sixteen thousand*: J. K. Blackshaw and C. Day, "Attitudes of Dog Owners to Neutering Pets: Demographic Data and Effects of Owner Attitudes," *Australian Veterinary Journal* 71, no. 4 (1994): 113–16.

135 *one male customer reported*: Bonnie Berry, "Interactionism and Animal Aesthetics: A Theory of Reflected Social Power," *Society and Animals* 16 (2008): 75–89.

135 *"many men continue to view . . ."*: Julie Urbanik, "'Hooters for Neuters': Sexist or Transgressive Animal Advocacy Campaign?" *Humanimalia* 1, no. 1 (2009): 40–62.

135 *Take tail docking*: Tom L. Beauchamp et al., "Cosmetic Surgery for Dogs," in *The Human Use of Animals: Case Studies in Ethical Choice*, 2nd ed. (New York: Oxford University Press, 2008), 135–46.

135 *The American Kennel Club (AKC), which develops*: The standards for each breed, including those listed in the text, are available at the AKC's website: "Breeds," American Kennel Club, accessed March 19, 2012, www.akc.org/breeds.

135*n* *For women, however*: "Merchandise Mart," CTI Neuticles, Novemeber 2, 2011, www.neuticles.com/shop/merchandisemart.shtml.

136 *Pet plastic surgey can have . . . crooked teeth*: Bonnie Berry, "Interactionism and Animal Aesthetics: A Theory of Reflected Social Power," *Society and Animals* 16 (2008): 75–89; Sandy Robins, "More Pets Getting Nipped and Tucked," MSNBC.com, April 27, 2005, www.msnbc.msn.com/id/6915955/ns/health-pet_health/t/more-pets-getting-nipped-tucked/<#>.T2e__syJnZR; James Hall, "Surge in Plastic Surgery for Pets," *Telegraph*, August 16, 2011, www.telegraph.co.uk/news/uknews/8704485/Surge-in-plastic-surgery-for-pets.html.

136 *But one veterinary surgeon*: Steve Kingstone, "Brazilian Dogs Go Under the Knife," BBC News, August 16, 2004, http://news.bbc.co.uk/2/hi/americas/3923099.stm.

136*n* *The United Kingdom has banned*: "Mutilations and Tail Docking of Dogs," Department for Environment Food and Rural Affairs, accessed March 12, 2012, http://archive.defra.gov.uk/foodfarm/farmanimal/welfare/act/secondary-legis/docking.htm.

137 *"We try to work . . ."*: Carroll, discussion, June 2010.

137 *Sandhill cranes commonly come*: Information about sandhill cranes, and the prostheses that Fox and Carroll built for them, are from Lee Fox, in discussion with author, Sarasota, Florida, March 26, 2011.

137 *"truly, physically sick . . ."*: Ibid.

137 *"Chrisie walked in hers . . ."*: Ibid.

138 *"Animals are wild . . ."*: Carroll, discussion, June 2010.

138 *A corgi, for instance*: Noel Fitzpatrick, in discussion with author via telephone, March 25, 2011.

138 *"Animals deserve a good quality . . ."*: Ibid.

138 *Feline and canine anatomy*: The limitations of regular prosthetics for dogs and cats comes from interviews with two sources: Fitzpatrick, discussion; and Denis Marcellin-Little, in discussion with author via telephone, March 22 and June 7, 2011.

139 *"a bionic dog"*: Ibid.

139 *Fitzpatrick knew that osseointegration*: Ibid.

139 *Blunn believed that surgeons*: Details about deer antlers, and their relevance to prosthetics, come from Gordon Blunn, in discussion with author via telephone, April 15, 2011; and C. J. Pendegrass et al., "Nature's Answer to Breaching the Skin Barrier: An Innovative Development for Amputees," *Journal of Anatomy* 209 (2006): 59–67.

139 *Using antlers as inspiration*: Fitzpatrick, discussion; Blunn, discussion.

139 *One of the first patients*: Information about Coal, and the ITAP procedure, comes from several sources: Fitzpatrick, discussion; Noel Fitzpatrick et al., "Intraosseous Transcutaneous Amputation Prosthesis (ITAP) for Limb Salvage in 4 Dogs," *Veterinary Surgery* 40, no. 8 (2011): 909–25; Noel Fitzpatrick, "Intraosseous Transcutaneous Amputation Prosthesis: An Alternative to Limb Amputation in Dogs and Cats," *Society of Practising Veterinary Surgeons Review* 2009 (2009): 43–46; "Coal's Story," Fitzpatrick Referrals, accessed March 20, 2012, www.fitzpatrickreferrals.co.uk/pet-owners/case-studies/coals-itap.

139*n* *"I watched it float away"*: Fitzpatrick, discussion.

140 *"The ITAP didn't just . . ."*: "Coal's Story," Fitzpatrick Referrals.

140 *a black cat named Oscar*: Liz Thomas, "Oscar the Bionic Cat," *Daily Mail*, June 25, 2010, www.dailymail.co.uk/sciencetech/article-1289281/Oscar-bionic-cat-pioneering-surgery-gave-TWO-false-legs.html; Adam Hadhazy, "Bionic Devices Let Injured Animals Roam Again," *Live Science*, July 15, 2010, www.livescience.com/10742-bionic-devices-injured-animals-roam.html.

140 *two dozen of these*: Fitzpatrick, discussion.

140 *The procedure has been*: Blunn, discussion.

140 *One of the first recipients*: Fitzpatrick et al., "Intraosseous Transcutaneous Amputation Prosthesis (ITAP)"; Norbert V. Kang et al., "Osseocutaneous Integration of an Intraosseous Transcutaneous Amputation Prosthesis Implant Used for Reconstruction of a Transhumeral Amputee: Case Report," *Journal of Hand Surgery* 35, no. 7 (2010): 1130–34.

140 *Denis Marcellin-Little*: Details about his patients and work come from Marcellin-Little, discussions, March and June 2011.

141 *Meanwhile, one equine researcher*: The researcher is Gary Sod at Louisiana State University. Information comes from Sod, in discussion with author via telephone, March 10, April 18, and June 6, 2011.

141 *Veterinarians tried to repair*: "Barbaro Euthanized After Lengthy Battle," MSNBC.com, accessed June 14, 2012, http://nbcsports.msnbc.com/id/16846723/ns/sports-horse_racing.

141 *"It's about life and love . . ."*: Fitzpatrick, discussion.

141 *Monkeys outfitted with brain implants*: Velliste et al., "Cortical Control of a Prosthetic Arm for Self-feeding"; Carmena et al., "Learning to Control a Brain-Machine Interface for Reaching and Grasping by Primates."

141 *paralyzed humans have performed*: "Paralyzed Man Uses Mind-Powered Robot Arm to Touch," *USA Today*, October 10, 2011, http://yourlife.usatoday.com/health/story/2011-10-10/Paralyzed-man-uses-mind-powered-robot-arm-to-touch/50716800/1.

141 *Scientists at the Rehabilitation Institute*: Todd A. Kuiken et al., "Targeted Muscle Reinnervation for Real-time Myoelectric Control of Multifunction Artificial Arms," *Journal of the American Medical Association* 301, no. 6 (2009): 619–28.

142 *In particular, the Defense*: "Revolutionizing Prosthetics," DARPA, accessed March 20, 2012, www.darpa.mil/Our_Work/DSO/Programs/Revolutionizing_Prosthetics.aspx; "DARPA's Revolutionizing Prosthetics Program Approaches Milestones," DARPA, October 10, 2011, www.darpa.mil/NewsEvents/Releases/2011/10/10.aspx.

7. Robo Revolution

143 *In the 1960s, the Central*: Information on Acoustic Kitty is from Robert Wallace and H. Keith Melton, *Spycraft: The Secret History of the CIA's Spytechs*

from Communism to al-Qaeda (New York: Dutton, 2008): 200–202; Jeffrey T. Richelson, *The Wizards of Langley: Inside the CIA's Directorate of Science and Technology* (Cambridge, MA: Perseus Books, 2001), 147–48; Charlotte Edwardes, "CIA Recruited Cat to Bug Russians," *Telegraph*, November 4, 2001, www.telegraph.co.uk/news/worldnews/northamerica/usa/1361462/CIA-recruited-cat-to-bug-Russians.html; and Julian Borger, "Project Acoustic Kitty," *Guardian*, September 11, 2001, www.guardian.co.uk/world/2001/sep/11/worlddispatch.

143 *as a heavily redacted CIA memo*: "Views on Trained Cats [redacted] for [redacted] Use," memorandum, March 1967, available at www.gwu.edu/~nsarchiv/NSAEBB/NSAEBB54/.

144 *In 2006, for example, DARPA*: Defense Advanced Research Projects Agency, *Hybrid Insect Micro Systems: Proposer Information Pamphlet* (BAA 06-22, March 9, 2006), available for download at https://www.fbo.gov/index?s=opportunity&mode=form&id=ec6d6847537a9220810f4282eedda0d2&tab=core&_cview=1.

144 *Building these machines . . . aloft for very long*: Michel Maharbiz, in discussion with author via telephone, February 8, 2010; Michel M. Maharbiz and Hirotaka Sato, "Cyborg Beetles," *Scientific American*, December 2010, 94–99.

144 *The Nano Hummingbird*: "Nano Hummingbird," AeroVironment, Inc., accessed April 23, 2012, www.avinc.com/nano; "AeroVironment Develops World's First Fully Operational Life-Size Hummingbird-Like Unmanned Aircraft for DARPA," AeroVironment, Inc., February 17, 2011, www.avinc.com/resources/press_release/aerovironment_develops_worlds_first_fully_operational_life-size_hummingbird; Dana Mackenzie, "It's a Bird, It's a Plane, It's a . . . Spy?" *Science* 335 (March 23, 2012): 1433.

144 *the DelFly Micro*: "DelFly Micro," DelFly, accessed April 26, 2012, www.delfly.nl/?site=diii&menu=home&lang=en; G. C. H. E. de Croon et al., "Design, Aerodynamics, and Vision-Based Control of the DelFly," *International Journal of Micro Air Vehicles* 1, no. 2 (2009): 71–97; Maharbiz and Sato, "Cyborg Beetles."

144 *"Proof-of-existence of small-scale . . ."*: Defense Advanced Research Projects Agency, *Hybrid Insect Micro Systems.*

144 *So far, nature's creations*: The advantages of insects are discussed in many papers, including Hirotaka Sato and Michel M. Maharbiz, "Recent Developments in the Remote Radio Control of Insect Flight," *Frontiers in Neuroscience* 4 (December 2010); Ethem Erkan Aktakka et al., "Energy Scavenging from Insect Flight," *Journal of Micromechanics and Microengineering* 21 (2011): 095016; Alper Bozkurt, "Balloon-Assisted Flight of Radio-Controlled Insect Biobots," *IEEE Transactions of Biomedical Engineering* 56, no. 9 (2009): 2304–2307.

145 *"it might be possible to . . ."*: Defense Advanced Research Projects Agency, *Hybrid Insect Micro Systems.*

145 *What the agency wanted . . . buildings or caves*: Ibid.

146 *"What I wanted at the end . . ."*: Maharbiz, discussion, February 2010.

146 *He figured that most scientists . . . beetle flight*: Ibid.; Maharbiz and Sato, "Cyborg Beetles."

146–47 *Maharbiz and his team . . . on the forehead*: Hirotaka Sato et al., "Remote Radio Control of Insect Flight," *Frontiers in Integrative Neuroscience* 3 (October 2009): article 24; Maharbiz and Sato, "Cyborg Beetles."

147 *Through trial and error . . . fall out of the air*: Michel Maharbiz, in discussion with author by telephone, January 4, 2012; Sato et al., "Remote Radio Control of Insect Flight"; Maharbiz and Sato, "Cyborg Beetles."

147 *The flower beetle's transformation began*: The steps required to turn a beetle into a flying machine are from Sato et al., "Remote Radio Control of Insect Flight"; Maharbiz and Sato, "Cyborg Beetles"; Sato and Maharbiz, "Recent Developments in the Remote Radio Control of Insect Flight"; and Maharbiz, discussion, January 2012.

148 *Then it was time . . . across the tile floor*: You can see these flights in the many videos Maharbiz and his students shot. They are posted alongside their 2009 paper (Sato et al., "Remote Radio Control of Insect Flight") at www.frontiersin.org/integrative_neuroscience/10.3389/neuro.07.024.2009/abstract.

148 *"The creation of a cyborg . . ."*: Sharon Weinberger, "Video: Pentagon's Cyborg Beetle Takes Flight," *Wired*, September 24, 2009, www.wired.com/dangerroom/2009/09/video-cyborg-beetle-takes-flight/.

148 *"Spies may soon be bugging . . ."*: "U.S. Military Create Live Remote-Controlled Beetles to Bug Conversations," *Daily Mail*, October 19, 2009, www.dailymail.co.uk/sciencetech/article-1221438/Ssh–conversation-bugged-cyborg-beetle.html<#>ixzz1ic22geJM.

148 *A columnist speculated*: Tracy Staedter, "Cyborg Beetles Employed as Military Weapons," *Discovery News*, November 18, 2009, http://news.discovery.com/tech/cyborg-beetles-employed-as-military-weapons.html.

148 *beetles that had been "zombified"*: Stuart Fox, "Video: DARPA's Remote-Controlled Cyborg Beetle Takes Flight," *Popular Science*, September 24, 2009, www.popsci.com/node/38759.

148 *references to "the impending robots . . ."*: Ross Miller, "Cyborg Beetles Commandeered for Test Flight, Laser Beams Not (Yet) Included," *Engadget*, January 29, 2009, www.engadget.com/2009/01/29/cyborg-beetles-commandeered-for-test-flight-laser-beams-not-ye/.

148 *When Maharbiz reflects upon*: Michel Maharbiz, in discussion with author, Berkeley, California, April 5, 2011.

149 *"Insects have inherently some sort . . ."*: Ibid.

149 *"some evil government conspiracy"*: Maharbiz, discussion, February 2010.

149 *"I think that's nonsense"*: Maharbiz, discussion, April 2011.

149 *"equally reprehensible"*: Maharbiz, discussion, February 2010.

149 *Imagine, Maharbiz tells me . . . search for survivors*: Ibid.

149 *"The fly is so small . . ."*: Maharbiz, discussion, April 2011.

149 *A Chinese research team*: Li Bao et al., "Flight Control of Tethered Honeybees

Using Neural Electrical Stimulation," *International IEEE EMBS Conference on Neural Engineering* (2011): 558–61.

149 *and Amit Lal, the engineer*: Denis C. Daly et al., "A Pulsed UWB Receiver SoC for Insect Motion Control," *IEEE Journal of Solid-State Circuits* 45, no. 1 (2010): 153–66; W. M. Tsang et al., "Insect-Machine Interface: A Carbon Nanotube-Enhanced Flexible Neural Probe," *Journal of Neuroscience Methods* 204, no. 2 (2012): 355–65; Alper Bozkurt, "Balloon-Assisted Flight of Radio-Controlled Insect Biobots," *IEEE Transactions of Biomedical Engineering* 56, no. 9 (2009): 2304–2307.

149–50 *One of Lal's innovations . . . around the implant*: Alper Bozkurt et al., "Insect-Machine Interface Based Neurocybernetics," *IEEE Transactions of Biomedical Engineering* 56, no. 6 (2009): 1727–33; Defense Advanced Research Projects Agency, *Hybrid Insect Micro Systems*; Sato and Maharbiz, "Recent Developments in the Remote Radio Control of Insect Flight."

150 *In one set of experiments*: Bozkurt, "Balloon-Assisted Flight of Radio-Controlled Insect Biobots."

150 *These kinds of pupal surgeries*: Advantages of implanting electronics in a pupa, rather than an adult insect, are discussed in Bozkurt, "Balloon-Assisted Flight of Radio-Controlled Insect Biobots"; and Bozkurt et al., "Insect-Machine Interface Based Neurocybernetics."

151 *"mass production of these . . ."*: Bozkurt, "Balloon-Assisted Flight of Radio-Controlled Insect Biobots."

151 *Our directional control . . . precisely 35 degrees*: Maharbiz, discussions, February 2010 and April 2011.

151 *navigate a complicated three-dimensional*: Maharbiz and Sato, "Cyborg Beetles."

151 *In 2011, a team of researchers*: Aktakka et al., "Energy Scavenging from Insect Flight."

152 *Usually, however, this work . . . new scientific feats*: S. K. Talwar et al., "Rat Navigation Guided by Remote Control," *Nature* 417, no. 6884 (May 2, 2002): 37–38; Shaohua Xu et al., "A Multi-channel Telemetry System for Brain Microstimulation in Freely Roaming Animals," *Journal of Neuroscience Methods* 133, nos. 1–2 (2004): 57–63.

152 *Rats have an excellent . . . cyborg insects*: Linda Hermer-Vazquez, in discussion with author via telephone, January 13, 2010.

152 *"They could fit through . . ."*: Ibid.

152 *They began by opening up*: Details about how the rats were created are from Talwar et al., "Rat Navigation Guided by Remote Control"; Xu et al., "A Multichannel Telemetry System"; John K. Chapin et al., Method and Apparatus for Guiding Movement of a Freely Roaming Animal Through Brain Stimulation, US Patent 7970476, filed February 10, 2003, issued June 28, 2011; John K. Chapin et al., Method and Apparatus for Teleoperation, Guidance, and Odor Detection Training of a Freely Roaming Animal Through Brain Stimulation, US Patent application no. 11/547,932, filed April 6, 2005, publication no. US 2009/0044761 A1; Hermer-Vazquez, discussion.

153–54 *Over the course of ten . . . steep ramp*: Talwar et al., "Rat Navigation Guided by Remote Control."

154 *As a final demonstration . . . conventional food rewards*: Linda Hermer-Vazquez et al., "Rapid Learning and Flexible Memory in 'Habit' Tasks in Rats Trained with Brain Stimulation Reward," *Physiology and Behavior* 84 (2005): 753–59; John K. Chapin et al., Method and Apparatus for Teleoperation; Hermer-Vazquez, discussion.

154 *"The robo-rats were . . ."*: Hermer-Vazquez, discussion.

154 *As Maharbiz wrote in an account*: Maharbiz and Sato, "Cyborg Beetles."

154 *Many animal liberationists*: Herzog, *Some We Love.*

155 *In fact, most Americans take*: Herzog, in discussion with author via telephone, November 4, 2011; Herzog, *Some We Love.*

155 *"you can end up . . ."*: Herzog, discussion.

155 *Herzog has found . . . cure for cancer*: Herzog, discussion; Herzog, *Some We Love.*

156 *Maharbiz notes that his beetles*: Maharbiz and Sato, "Cyborg Beetles."

156 *an actual suggestion the SUNY researchers*: John K. Chapin et al., Method and Apparatus for Guiding Movement.

156 *"Where do you draw . . ."*: Maharbiz, discussion, April 2011.

157 *"Maybe I'm an example . . ."*: Maharbiz, discussion, February 2010.

157 *That's fine with Maharbiz*: Ibid.

157 *"to get people to think . . ."*: Maharbiz, discussion, April 2011.

157 *The technique, which comes*: General information about optogenetics techniques and how they work is from a number of sources, including Ed Boyden, in discussion with author via telephone, September 9, 2011, and January 12, 2012; Edward S. Boyden et al., "Millisecond-Timescale, Genetically Targeted Optical Control of Neural Activity," *Nature Neuroscience* 8 (2005): 1263–68; Edward S. Boyden, "A History of Optogenetics: The Development of Tools for Controlling Brain Circuits with Light," *F1000 Biology Reports* 3 (May 2011); Karl Deisseroth, "Controlling the Brain with Light," *Scientific American*, October 20, 2010, www.scientificamerican.com/article.cfm?id=optogenetics-controlling; editorial, "Enlightened Engineering," *Nature Biotechnology* 29 (October 13, 2011): 849.

158 *By turning certain neurons on and off*: Boyden, discussion, 2011; Tomomi Tsunematsu et al., "Acute Optogenetic Silencing of Orexin/Hypocretin Neurons Induces Slow-Wave Sleep in Mice," *Journal of Neuroscience* 31, no. 29 (July 2011): 10529–39.

158 *Or we can use a beam of light*: Dayu Lin et al., "Functional Identification of an Aggression Locus in the Mouse Hypothalamus," *Nature* 470 (February 10, 2011): 221–26.

158 *In 2011, Edward Boyden*: Information about the wireless helmet study is from Boyden, discussions, 2011 and 2012; Christian T. Wentz et al., "A Wirelessly Powered and Controlled Device for Optical Neural Control of Freely-Behaving Animals," *Journal of Neural Engineering* 8, no. 4 (2011).

159 *"It's sort of turning up . . ."*: Boyden, discussion, 2012.

159 *To Boyden, the headset*: Ibid. For more on potential clinical applications, see "Enlightened Engineering," *Nature Biotechnology* 29.

159–60 *a pair of former neuroscience postdocs . . . a neuron fire*: Greg Gage and Tim Marzullo, in discussion with author, Woods Hole, Massachusetts, August 22–23, 2011.

160 *In 2009, Gage and Marzullo . . . with their students*: Ibid. For more on the SpikerBox, and how it can be used in the classroom, see: Timothy C. Marzullo and Gregory J. Gage, "The SpikerBox: A Low Cost, Open-Source BioAmplifier Increasing Public Participation in Neuroscience Inquiry," *PLoS One* 7, no. 3 (2012): e30837.

160*n* *The Backyard Brains website*: "Spike Counter," Backyard Brains, accessed June 15, 2012, www.backyardbrains.com/SpikeCounter.aspx.

161 *Because a roach relies*: Gage and Marzullo, discussion.

161 *"It's like* designed *to be . . ."*: Tim Marzullo, in discussion with author, Woods Hole, Massachusetts, August 23, 2011.

161 *Marzullo has spent the morning*: Gage and Marzullo, discussion.

161 *"We actually don't know . . ."*: "Ethical Issues Regarding the Use of Invertebrates in Education," Backyard Brains, accessed June 9, 2012, http://wiki.backyardbrains.com/Ethical_Issues_Regarding_the_Use_of_Invertebrates_in_Education.

161 *Steering the roach*: Gage and Marzullo, discussion.

162 *The electronics are slightly modified*: Ibid.

163 *Nevertheless, the RoboRoach*: Ibid.

163 *as of June 2012*: Tim Marzullo, e-mail message to author, June 4, 2012.

163 *"It's kind of edgy . . ."*: Marzullo, discussion.

163 *"We're doing all this . . ."*: Ibid.

164 *"But if you exploit animals . . ."*: Ibid.

164 *there is a growing community*: For more on biohackers, see Delthia Ricks, "Dawn of the Biohackers," *Discover*, October 2011, http://discovermagazine.com/2011/oct/21-dawn-of-the-biohackers; Erin Biba, "Genome at Home: Biohackers Build Their Own Labs," *Wired*, September 2011, www.wired.com/magazine/2011/08/mf_diylab/all/1; Ritchie S. King, "When Breakthroughs Begin at Home," *New York Times*, January 16, 2012, www.nytimes.com/2012/01/17/science/for-bio-hackers-lab-work-often-begins-at-home.html; Pui-Wing Tam, "'Biohackers' Get Their Own Space to Create," *Wall Street Journal*, January 12, 2012, http://online.wsj.com/article/SB1000 1424052970204124204577150801888929704.html; and "DIYbio," DIYbio, accessed April 26, 2012, http://diybio.org/.

164*n* *The pair rarely lets*: Gage and Marzullo, discussion.

165 *As it happens, their most*: "Backyard Brains Returns to the Nature Neuroscience Podcast, Unveils Optogenetics Prototype," Backyard Brains, accessed January 13, 2012, http://news.backyardbrains.com/?p=962; Gage and Marzullo, discussion.

165 *A class of New York*: "High School Students Hack Our RoboRoach Kit, Make It Better," Backyard Brains, accessed January 13, 2012, http://news.backyard brains.com/2011/08/high-school-students-hack-our-roboroach-kit-make-it -better/.

165 *Another customer—a Microsoft programmer*: Gage and Marzullo, discussion.

165 *We already have the annual International Genetically*: "The iGEM Foundation," The iGEM Foundation, accessed April 26, 2012, http://igem.org/About.

165 *In past years, teams*: "Team Groningen," iGEM 2009, accessed April 26, 2012, http://2009.igem.org/Team:Groningen; "Team Cambridge," iGEM 2009, accessed April 26, 2012, http://2009.igem.org/Team:Cambridge; Emily Singer, "Bizarre Bacterial Creations," *Technology Review*, November 3, 2006.

165 *"kids will be able to . . ."*: Maharbiz, discussion, April 2011.

8. Beauty in the Beasts

167 *there is a movement*: For more, see "Great Ape Project," Project GAP, www .greatapeproject.org/.

167 *In December 2011, the National*: "Statement by NIH Director Dr. Francis Collins on the Institute of Medicine Report Addressing the Scientific Need for the Use of Chimpanzees in Research," NIH News, National Institutes of Health, December 15, 2011, www.nih.gov/news/health/dec2011/od-15.htm.

167 *a bill currently before Congress*: Great Ape Protection and Cost Savings Act of 2011, H.R. 1513, 112th Cong. (2011).

167 *A number of other nations . . . effect in 2013*: 289. For more details on some of these bans see "International Bans," New England Anti-Vivisection Society, accessed November 13, 2012, www.releasechimps.org/laws/international-bans. On the protection of animals used for scientific purposes, Directive 2010/63/ EU, European Parliament (2010).

167 *Some American cities have passed*: "Guardian Communities," Guardian Campaign, accessed March 13, 2012, www.guardiancampaign.com/guardiancity .html.

167–68 *Seventy percent of dog owners*: Schaffer, *One Nation Under Dog*, 18.

168 *Americans spend $48 billion . . . eating animal flesh*: Hal Herzog, "Are We Really a Nation of Animal Lovers?" *Animals and Us* (blog), *Psychology Today*, February 14, 2011, www.psychologytoday.com/blog/animals-and-us/201102 /are-we-really-nation-animal-lovers.

168 *When Harold Herzog . . . surveyed*: Herzog, *Some We Love*, 239–40.

168 *In a Gallup poll*: David W. Moore, "Public Lukewarm on Animal Rights," Gallup News Service, May 21, 2003, www.gallup.com/poll/8461/public-lukewarm -animal-rights.aspx.

168 *These conflicting attitudes*: Herzog, *Some We Love*, 11–12.

168 *"Some argue that we are . . ."*: Ibid., 12.

169 *"I know that physiology . . ."*: Charles Darwin, "Mr. Darwin on Vivisection," *Times* (London), April 18, 1881.

169 *"We've always had strong . . ."*: Twine, discussion, 2009.

170 *One analysis of fifty . . . hereditary afflictions*: L. Asher et al., "Inherited Defects in Pedigree Dogs. Part 1: Disorders Related to Breed Standards," *The Veterinary Journal* 182 (2009): 402–11.

170 *Dalmatians are prone to . . . terrible hips*: "Inherited Diseases in Dogs Database," University of Cambridge, accessed March 7, 2012, www.vet.cam.ac.uk /idid/.

170 *As of 2012, commercial labs*: Cathryn Mellersh, "DNA Testing and Domestic Dogs," *Mammalian Genome* 23 (2012): 109–23.

170 *For less than a hundred dollars*: "VetGen—Veterinary Genetic Services," VetGen, accessed March 8, 2012, www.vetgen.com.

170*n* *For example, Cavalier King Charles spaniels*: Asher et al., "Inherited Defects in Pedigree Dogs. Part 1"; Nicola Rooney and David Sargan, "Pedigree Dog Breeding in the UK: A Major Welfare Concern?" (UK: Royal Society for the Prevention of Cruelty to Animals, 2009); Companion Animal Welfare Council, *Breeding and Welfare.*

171 *genetic testing, followed by . . . and corgis*: Rooney and Sargan, "Pedigree Dog Breeding in the UK."

171 *gene therapy experiments have been*: Glenn P. Niemeyer, "Long-term Correction of Inhibitor-Prone Hemophilia B Dogs Treated with Liver-Directed AAV2-Mediated Factor IX Gene Therapy," *Blood* 113, no. 4 (2009): 797–806; Katherine Parker Ponder et al., "Therapeutic Neonatal Hepatic Gene Therapy in Mucopolysaccharidosis VII Dogs," *PNAS* 99, no. 20 (2002): 13102–13107; S. J. M. Niessen et al., "Novel Diabetes Mellitus Treatment: Mature Canine Insulin Production by Canine Striated Muscle Through Gene Therapy," *Domestic Animal Endocrinology* (available online February 21, 2012).

171 *These dogs had all been*: Gustavo Aguirre, in discussion with author via telephone, April 5, 2012; Gregory M. Acland et al., "Gene Therapy Restores Vision in a Canine Model of Childhood Blindness," *Nature Genetics* 28 (May 2001): 92–95; "RPE65," U.S. National Library of Medicine, National Institutes of Health, accessed March 12, 2012, http://ghr.nlm.nih.gov/gene/RPE65.

171 *In 2001, Gustavo Aguirre*: Details of the gene therapy trial are from Acland et al., "Gene Therapy Restores Vision"; and Aguirre, discussion.

172 *And the fix was permanent*: Aguirre, discussion.

172 *For blind animals—and humans*: There are many papers and ongoing projects in this field. Here are a few: James D. Weiland, et al., "Retinal Prostheses: Current Clinical Results and Future Needs," *Ophthalmology* 11, no. 118 (2011): 2227–37. Gerald J. Chader, et al., "Artificial Vision: Needs, Functioning, and Testing of a Retinal Electronic Prosthesis," *Progress in Brain Research* 175 (2009): 317–32. J. D. Loudin, et al., "Optoelectronic Retinal Prosthesis: System Design and Performance," *Journal of Neural Engineering* 4 (2007): S72–84.

172 *Helen Sang, a developmental . . . feathered flocks*: Sang, discussion; Jon Lyall et al., "Suppression of Avian Influenza Transmission in Genetically Modified Chickens," *Science* 331 (January 14, 2011): 223–26.

172 *"What I would like to see happen . . ."*: Kraemer, discussion, December 2010.

172 *"You have to bear in mind . . ."*: Twine, discussion, 2012.

173 *Or consider a more extreme*: "Not Grass-Fed, but at Least Pain-Free," *New York Times*, February 19, 2010.

174 *George Dvorsky, a Canadian bioethicist . . . same technologies*: George Dvorsky, in discussion with author via telephone, February 15, 2010, and March 16, 2012; Geroge Dvorsky, "All Together Now: Developmental and Ethical Considerations for Biologically Uplifting Nonhuman Animals," *Journal of Evolution and Technology* 18, no. 1 (2008): 129–42.

175 *"Their horizon line is extremely . . ."*: Dvorsky, discussion, 2010.

175 *Dvorsky also imagines . . . complex forms of language*: Ibid.

175 *"I realize how absolutely . . ."*: Ibid.

175 *Scientists have already engineered*: Ya-Ping Tang et al., "Genetic Enhancement of Learning and Memory in Mice," *Nature* 401 (September 2, 1999): 63–69; Jonah Lehrer, "Small, Furry . . . and Smart," *Nature* 461 (October 14, 2009): 862–64.

175 *Another team of researchers*: Theodore W. Berger et al., "A Cortical Neural Prosthesis for Restoring and Enhancing Memory," *Journal of Neural Engineering* 8, no. 4 (2011): 046017.

175 *"blurring of the species line"*: Dvorsky, discussion, 2012.

175 *"the entire biosphere"*: Dvorsky, discussion, 2010.

176 *as the planet warms, birds*: *Birds and Climate Change: Ecological Disruption in Motion* (National Audubon Institute, 2009).

176 *The United Nations Intergovernmental Panel*: Intergovernmental Panel on Climate Change, *Climate Change 2007* (Geneva, Switzerland: United Nations, 2007).

176 *Though a big ram . . . horn sizes*: Chris T. Darimont et al., "Human Predators Outpace Other Agents of Trait Change in the Wild," *PNAS* 106, no. 3 (2009): 952–54; David W. Coltman et al., "Undesirable Evolutionary Consequences of Trophy Hunting," *Nature* 426 (December 11, 2003): 655–58.

176 *Similarly, fish have adapted*: Stephen Palumbi, "Humans as the World's Greatest Evolutionary Force," *Science* 293 (September 7, 2001): 1786–90.

177 *"I'm of the persuasion . . ."*: Kraemer, discussion, October 2009.

178 *If the world's farms*: Digital Angel Corp. patented the Bio-Thermo Microchip in 2006. (Vincent K. Chan and Ezequiel Mejia, "Method and Apparatus for Sensing and Transmitting a Body Characteristic of a Host," US Patent 7015826, filed April 2, 2002, issued March 21, 2006.) When the chips were introduced, the president of Digital Angel said the company planned to target chicken farmers and producers (Ephraim Schwartz, "Could Chips in Chickens Track Avian Flu?" *PC World*, December 6, 2005, www.pcworld.com/article/123845/could_chips_in_chickens_track_avian_flu.html; "Poultry Microchip on Watch for Bird Flu," UPI, December 5, 2005, www.upi.com/Science_News/2005/12/05/Poultry-microchip-on-watch-for-bird-flu/UPI-82541133811677/). However, Digital Angel sold Destron Fearing, its animal

ID unit, in 2011 ("Digital Angel Closes Sale of Destron Fearing Unit," Digital Angel, accessed March 7, 2012, www.digitalangel.com/presspost.php?passedcount=4). The current Destron Fearing website (www.destronfearing.com/, accessed March 7, 2012) does not mention the use of the chips in chickens.

179 *In 2006, scientists at Case Western*: Richard W. Hanson and Parvin Hakimi, "Born to Run: The Story of the PEPCK-C^{mus} Mouse," *Biochimie* 90, no. 6 (2008): 838–42.

180 *Some of the vision disorders . . . visually impaired people*: Aguirre, discussion.

180 *In 2012, for instance, a team of Swiss researchers*: Rubia van den Brand, et al, "Restoring Voluntary Control of Locomotion after Paralyzing Spinal Cord Injury," *Science* 336, no. 6085 (2012): 1182–85.

Acknowledgments

Writing a book is a solitary endeavor, but researching, reporting, polishing, and publishing one is anything but. Many people provided crucial assistance along the way.

First and foremost, thanks to the many scientists who invited me into their labs and lives. Many of them are mentioned in the preceding pages. Others do not appear in the book but provided me with invaluable background information and context. I am grateful to all of them. Without the generosity of these very busy researchers, this book literally would not have been possible.

I appreciate all those who read early chapter drafts and provided feedback. Nick Summers, Michelle Sipics, Blaine Boman, Alison Anthes, Gary Anthes, and Caroline Mayer, you have helped make this book better.

Thanks to the entire team at Scientific American / Farrar, Straus and Giroux. Amanda Moon has been a superlative editor throughout this process. Her energy and enthusiasm for this project equaled my own. Amanda's incisive comments and gentle suggestions helped turn my manuscript from a jumble of sentences into a coherent whole. Karen Maine became my trusty citations, formatting, and style guru, and Chris Richards ably took over where she left off. Kathy Daneman and the rest of the publicity and marketing brigade at FSG and *Scientific American* worked tirelessly to get the book into readers' hands.

I am indebted to everyone at the Park Literary Group, but especially to Theresa Park, who gave a young writer a chance, and Abigail Koons, who is the best agent a girl could ask for. Abby helped me navigate the transition from journalist to author and became my invaluable guide to the world of publishing. She provided spot-on writing and professional advice and became not just an agent, but also a motivator, cheerleader, and friend. I owe both my book and my sanity to her. (Also, special thanks to Blair Wilson for keeping Abby sane and the office running smoothly.)

On a more personal note, I'd like to thank my boyfriend, Blaine, who ably handled my occasional meltdowns; my sister, Ali, who kept me well-stocked with baked goods; and my parents, who provided more support than can possibly be detailed here. It's their own journalistic careers that introduced me to the field I have come to love, and for that—and so many other things—I will be forever grateful.

Finally, I must acknowledge Artemis, CC, Bruce, Dewey, Winter, Chrisie, Jonathan Sealwart, GloFish 1 through 6, Woods Hole 1 and 2, and all the other animals that gamely tolerated my intrusion into their lives. Though they did not volunteer to become guinea pigs or lightning rods, science and society owe these creatures one giant, collective thank-you note.

And, of course, lots of love to Milo, whose insatiable need for ear rubs and neighborhood walks kept my body from cramping up during the longest writing sessions. He is my own little beast.

Index

A NOTE ABOUT THE AUTHOR

Emily Anthes is a journalist whose articles have appeared in *Wired*, *Scientific American*, *Psychology Today*, *Slate*, *The Boston Globe*, and other publications. She holds a master's degree in science writing from MIT and a bachelor's degree in the history of science and medicine from Yale. She lives in Brooklyn, New York, with her dog, Milo. You can visit her website at www.emilyanthes.com.